Penguin Education
Penguin Library of Physical Sciences

Organic Mechanisms
R. Bolton

Advisory Editor
V. S. Griffiths

General Editors
Physics: N. Feather
Physical Chemistry: W. H. Lee
Inorganic Chemistry: A. K. Holliday
Organic Chemistry: G. H. Williams

Organic Mechanisms
R. Bolton

Penguin Books

Penguin Books Ltd, Harmondsworth, Middlesex, England
Penguin Books Inc, 7110 Ambassador Road, Baltimore, Md 21207, USA
Penguin Books Australia Ltd, Ringwood, Victoria, Australia

First published 1972
Copyright © R. Bolton, 1972

Made and printed in Great Britain by
William Clowes & Sons Ltd, London, Colchester and Beccles
Set in IBM Press Roman

Contents

Editorial Foreword

For some years, it has been obvious to most teachers of organic chemistry that the day is past when one author, or a small group of authors, could write a comprehensive textbook of organic chemistry suitable for the courses in this branch of the subject as taught in universities. While a number of excellent textbooks are available, they must now almost of necessity be either incomplete in their coverage of the subject or massively unwieldy and correspondingly expensive.

It seemed to the publishers and to the editor that a useful method of coping with this difficulty was to present the material as a series of smaller texts on various aspects of the subject; by producing them in the format used in this series their price has been kept down to a minimum. In the books on organic chemistry, those parts of the subject which are generally regarded as essential and common to all universities' degree courses will be covered, as also will many topics which appear as optional subjects in many degree courses. It is hoped that students will wish to purchase such of the volumes as are relevant to their own particular courses.

The books in the series have been planned essentially as undergraduate texts although it is also hoped that they will be of use to more advanced students who require concise introductions to particular aspects of the subject. Since each book has been written by experts in its subject, it is hoped that all the volumes will come to be regarded as authoritative texts. It is also hoped that undergraduates and other readers will appreciate the inclusion of some worked examples and problems in many of the books.

G.H.W.

Preface

Physical organic chemistry has its roots in the mid-nineteenth century. The vitalistic theories, which ascribed especial properties to organic molecules because of their animal or vegetable origin, were being discarded at this time, mainly as a result of the syntheses of urea (Wohler, 1821, 1828) and acetic acid (Kolbe, 1845) from 'inorganic' reagents. Once it was recognized that no essential differences existed between compounds containing carbon and those containing other elements it became meaningful to try to understand the structure and reactions of organic species in terms of the 'normal' forces and interactions of inorganic and physical chemistry.

Since physical and chemical knowledge had not advanced sufficiently to provide a basis for theory until the beginning of the present century, the real advances in physical organic chemistry came in the 1920s and 1930s. It is now possible to trace the course of many organic reactions more or less exactly, although there are a number of questions which are still unanswered. These unknowns may be dealt with in two ways.

A problem which cannot be fully understood by our present experimental techniques (e.g. the nature of bonding, the structure of the atom) may be dealt with by a theory. This theory must be consistent with the experimental facts but need not reflect the true state of affairs (as there may be insufficient evidence to decide between two models or explanations). Alternatively, we can provide a general explanation of the problem without a detailed theory (e.g. solvent polarity).

Ultimately all our thinking is based upon theory, and we should not forget the limitations of both our theories and our nomenclature.

Reaction mechanisms seek to define the course of a reaction exactly. None succeed; moreover, the evidence upon which they are based is often uncompelling and may even be ambiguous. Some proposals come from a detailed study of the reaction, in which *all* the evidence agrees with the proposed mechanism (and other plausible ones are eliminated, or at least their existence is recognized). Other processes have been suggested solely on one piece of evidence (e.g. the isolation of an 'unusual' reaction product), and more properly reflect the current passion for providing mechanistic explanations for every observation rather than an exhaustive research investigation.

Before accepting a proposed mechanism, one ought to test its basis and see how well it is founded. Since most reactions will proceed under a variety of conditions (e.g. chlorination or nitration of benzene, which occurs in the gas phase or in solution, and in the presence or absence of catalysts) no one mechanism will usually cover all the possible situations. It follows that we cannot describe a process as '*the* mechanism of the reaction'. Indeed, we can only properly invoke a published mechanism if we want to relate it to work done under identical experimental conditions. In other circumstances, although the experimental conditions are similar, it is wise to say, for example, that 'The reported mechanism of nitration of benzene in nitromethane under specified conditions is . . . and this may well prevail under the present conditions.'

In many of the reactions discussed in this book, the evidence for the proposed mechanism is considerable. In some cases, however, there is still discussion; the only way in which the student can reach his own conclusion on the merits of these arguments is by consulting the original papers. There has been neither the space nor the time to consider the details of every reaction known in organic chemistry; other texts should be consulted for individual mechanisms not dealt with here.

The purpose of this text is to introduce the student to the nomenclature of physical organic chemistry, and to try to show the logic upon which it is based. As in all other branches of science, we must say 'Is it true?' and, if we cannot recognize the truth, 'Is it logical?'. I hope that students will be encouraged to apply this critical faculty to the arguments outlined in the text. The questions at the end of each chapter may help here. They are meant to apply the principles of physical organic chemistry to other reactions; where possible, these have been taken from the literature.

Since individual literature references are not given, a short list of relevant review articles are included under each chapter heading in Chapter 13.

Finally, I must acknowledge defeat. In writing such a text, one hopes to improve upon previous works. However, the books of E. S. Gould and of J. Hine remain as examples of the best in physical organic chemistry, and I acknowledge my debt to them, which will be evident in the following pages!

Chapter 1
Theoretical Considerations

This chapter deals with atomic and molecular structure in order to show what molecules really are and to explain the theoretical basis of organic chemistry. Although much of the material is common to a number of texts, it is presented again to give a clear understanding of the various processes in physical organic chemistry.

1.1 Atomic structure

The present theory of the structure of matter has the atom as its basis; this is the smallest particle containing the identity of the element. A hydrogen atom, for instance, differs from a carbon atom. All atoms, however, consist of a nucleus containing protons, having a single positive charge, and generally neutrons having a similar mass to the proton but no charge. This nucleus is surrounded by electrons having unit negative charge and a much smaller mass than the proton. All electrons are equivalent in the sense that we cannot distinguish an electron of a copper atom from an electron of a sulphur atom. The same is true of protons and of neutrons.

Since the total charge of any atom is zero, there must be an equivalent number of protons and electrons. The electrons are arranged in *shells* corresponding to their energy; each shell can accommodate a fixed number of electrons only in a few well-defined energy levels. The lowest energy level, corresponding to the shell most readily filled, is that in which the electron is closest to the nucleus; usually, but *not* inevitably, the energy level of the shell increases with increasing distance from the nucleus.

1.1.1 *Quantum numbers*

To each shell is designated an integer n called the *principal quantum number; n* can have the values $1, 2, 3, \ldots$ etc. Letters are also used, the K-shell corresponding to $n = 1$, the L-shell to $n = 2$, etc. Within the shell there are a number of energy levels described by the *orbital quantum number l*, which has integral values between zero and $n - 1$, and by a third quantum number, m, the *magnetic quantum number*, which can have the values $-l, -l + 1, \ldots, -1, 0, 1, \ldots, l - 1, l$. A fourth number allows for a property of the electron called *spin*. In a magnetic or an electric field, two types of electrons can be distinguished. One type enhances the field (spins 'with' the field; i.e. $H + dH$ = total field) and the other opposes the field (spins 'against' the field; i.e. $H - dH$ = total field). The difference in behaviour can be understood by considering the electron as a small magnet able to spin either clockwise or anticlockwise. The spin number simply indicates that two types of electron exist; the number is usually termed $\frac{1}{2}$ or $-\frac{1}{2}$ although the fraction has no physical significance and we could just as well use the terms s and $-s$.

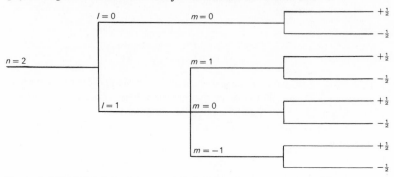

Figure 1 Energy levels in the second shell (L-shell)

These energy levels have some bearing upon the appearance of the electron cloud about the nucleus; n is related to the *size* of the cloud, l is linked to various *shapes* of the cloud, and m is associated with its *orientation* in space. This orientation must necessarily be with respect to some other cloud in the system, for there is no definite direction to which atoms point. Note that we have talked about 'clouds' and not orbits; this will be justified later on in the text.

1.1.2   *The* Aufbau *principle*

If the electrons are most stable in the lowest-energy states, and if the energy level increases going from the K-shell to the L-shell and so on, there is an obvious way in which the electronic configurations of the elements may be 'built up' (German: *aufbauen*). The first electron will go into the K-shell where, as there is only one value of l and of m, there is only one energy level. This corresponds to the hydrogen configuration. A second electron will lie in the same energy level, but antiparallel (with opposed spin) to the first, giving the helium configuration. Two electrons in the one energy level must be paired, or antiparallel, so that their overall energy is a minimum. No more than two electrons may be accommodated in any one level. The K-shell is now filled; there are no unfilled energy levels.

A third electron must now be placed in the higher-energy L-shell; as there are two values of l and three values of m, a total of eight electrons can be incorporated within the shell (Figure 1). Proceeding in this way, the M- and N-shells may be successively filled. This is the *Aufbau* principle.

The energy levels at $l = 0, 1, 2$ and 3 are also described by the letters s, p, d, and f; these terms arise from spectroscopic usage. With this terminology, the energy level of the K-shell is called 1s ($n = 1, l = 0$), those of the L-shell are called 2s ($n = 2, l = 0$) and 2p ($n = 2, l = 1$), and those of the M-shell are 3s ($l = 0$), 3p ($l = 1$) and 3d ($l = 2$).

1.3 *The nature of the electron and its atomic position*

Previously the term 'electron cloud' has been used, although the energy of the electron has been carefully defined. This implies (correctly) that the electron does not travel a regular orbit around the nucleus. The Heisenberg uncertainty principle explains why both the position and the energy of a particle such as an electron cannot be accurately defined. Electrons possess both mass and momentum, like particles, but can be diffracted, like beams of light. The Heisenberg principle generally underlines the overlapping nature of these two properties, so that the complete behaviour of the electron cannot be described solely in terms of particle or wave characteristics. For any species possessing both particle and wave character the uncertainty in describing these two properties is appreciable,

$$\Delta E \, \Delta x \simeq h, \hspace{4cm} \textbf{1.1}$$

where ΔE is the uncertainty in the value of the radiation energy E of the species, Δx the uncertainty in x (the coordinate which defines position) and h is Planck's constant. The error involved becomes appreciable only when dealing with sub-atomic physics ($h = 6 \cdot 6 \times 10^{-34}$ J s). By placing the electron in a specific energy level we have defined its energy to very small limits; its position is therefore uncertain and the error is, in fact, of the same order as the atomic radius.

It is conventional to draw a three-dimensional shape within which the electron is almost certain to be found at any time; this usually corresponds to 90 per cent, 95 per cent or 99 per cent probability. The shapes of these orbitals are similar, but their sizes differ. As this surface's shape and position depend upon the energy of the electron, this spatial representation is called an orbital as well. Alternatively, a plot of the distribution of the electron with time produces something of the same shape as the orbital, and looking rather like a cloud – it is from this representation that the term 'electron cloud' arises. The uncertainty principle rules out the concept of an electron fixed in an orbit around the nucleus. Since most of our discussion deals with elements of the first row of the periodic table, and particularly carbon, we shall not consider the details of d-, f- and other higher-energy orbitals.

1.4 *Orbital shape and hydridization*

The s-orbitals are spherical; in other words, it is equally probable that the electron will occur in any direction, but usually within a definite distance, from the nucleus. The p-orbitals are situated orthogonally (i.e. mutually at 90 °) and are therefore distinguished by the terms p_x, p_y and p_z. There is no rule governing which p-orbital is called p_x, since they are initially equivalent and are symmetrically distributed about the nucleus.

The carbon atom has the electronic structure $1s^2$, $2s^2$, $2p^2$; this implies two types of bond, one involving the 2s electrons and the other involving the higher-energy 2p electrons. In fact, all four bonds in a substance like methane, CH_4, are equivalent. However, in order to form a bond the atom must be activated* and it

* To occur, any reaction requires some energy; if it does not, the reagents are spontaneously unstable and cannot exist. This activation energy is quite distinct from the energy of the reaction.

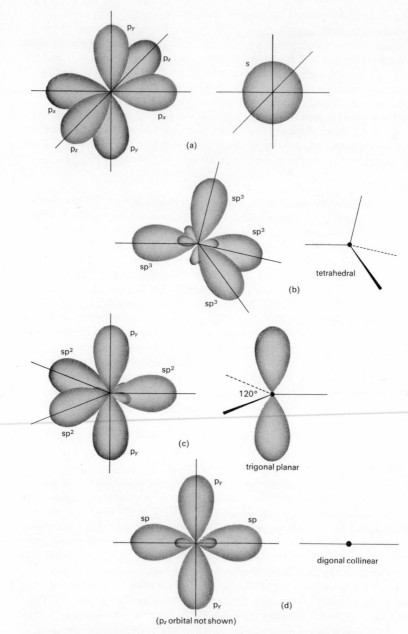

Figure 2 (a) Unhybridized s- and p-orbitals, (b) tetrahedral sp^3 hybrid, (c) sp^2 trigonal hybrids, (d) sp digonal hybrid

is presumed that under this activation the four electrons are firstly promoted to the states $2s^1$, $2p_x^1$, $2p_y^1$ and $2p_z^1$, and that these four levels then *hybridize* (i.e. combine together, losing their individual identities) to give four equivalent energy levels.

These four sp^3 orbitals are tetrahedrally arranged (Figure 2b) which is consistent with the known tetrahedral geometry of carbon bonding.

Other hybrids may be formed involving one s and two p electrons to give three equivalent sp^2 orbitals (Figure 2c) which are mutually at $120°$ and coplanar (trigonal). Another possible hybridization may occur between one s- and one p-orbital to give two sp hybrids (Figure 2d); these are collinear (digonal).

1.2 Bonding

The electronic structure of atoms has been described in terms of definite energy levels giving rise to orbitals of different sizes and shapes. If all atoms tend to the electronic structure of inert gases (when the shells are completely filled) some aspects of bonding may be simply explained.

1.2.1 *Ionic bonding*

The reaction of lithium with fluorine involves an electron being lost from the 2s orbital of lithium and filling the 2p orbitals in fluorine. The resulting ions now have the electronic configurations of helium and neon respectively, and the bond between the two oppositely charged species is an ionic bond. Similarly, other elements in Groups I and II especially can lose electrons to give ions with the inert gas structure, and other elements in Groups VI and VII accept electrons to give ions with completed outer shells of electrons. The two ions are held together by electrostatic forces. In a crystal of NaCl, for example, each sodium ion is surrounded by six chloride ions; it is not particularly bonded to any one of them (cf. covalent bond).

1.2.2 *Covalent bonding*

Ionic bonding seems unlikely in a species such as methane. In this case, four electrons would have to be lost or gained by carbon to achieve the inert-gas configuration, and the energies required to remove electrons from intermediate species such as C^{3+}, or to insert electrons into a species C^{3-}, would be ridiculously high. A similar situation occurs in the bonding of similar or identical atoms (e.g. IBr or H_2). G. N. Lewis proposed a *covalent* or *electron-pair* system of bonding in which one or more pairs of electrons were shared between two atoms and constituted a bond. By analogy with the ionic bond it was proposed that the covalent bond was most stable when each atom involved was associated with an outer shell of eight electrons (or one with the inert gas configuration). In this way the existence of stable substances such as methane, molecular hydrogen and iodine monobromide could be explained.

Covalently bonded substances are most easily recognized by their relatively low melting points and boiling points. In an ionic system there are electrostatic interactions between one anion and all its oppositely charged neighbours in the crystal; as a result, bonding can be considered as proceeding through the entire molecule and the energy needed to disorder such a system (i.e. melting) is high. In the case of covalently bonded substances only weak interactions occur between the molecules, and so disorder (melting and boiling) occurs relatively easily. Unlike the ionic bond, a covalent bond links two specific atoms in a definite direction.

1.2.3 Dative bonding

Since covalent bonding involves two electrons being shared by two atoms, it is a small step to consider that both of these electrons might be supplied by one atom to the other, so that both can now acquire the inert gas structure. Such bonding is termed *dative* or *coordinate* bonding, and occurs in the formation of NH_4^+ from NH_3 and H^+, and the formation of BF_4^- from BF_3 and F^-.

In the first case, it is the unshared pair of electrons on nitrogen which is donated; in the second example, the fluoride ion provides the electrons to boron.

1.2.4 Bonding in carbon compounds

Carbon compounds usually result from covalent bonding. The overlapping of two atomic orbitals which is essential if the *two* bonding electrons can be shared by the two bonded atoms, forms a *molecular orbital* within which the two electrons constituting the bonding electrons may be found, more probably in the region of overlap. Between two carbon atoms a *sigma* (σ-) bond results from two overlapping hybridized orbitals and their associated electrons. Two unhybridized p-orbitals

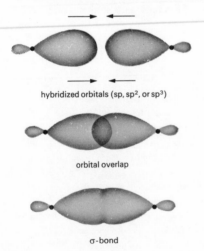

hybridized orbitals (sp, sp², or sp³)

orbital overlap

σ-bond

Figure 3a Bonding between two carbon atoms: σ-bonding

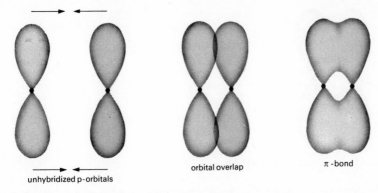

unhybridized p-orbitals   orbital overlap   π-bond

Figure 3b Bonding between two carbon atoms: π-bonding

may overlap to form a *pi* (π-) bond. These two bonds have different characteristics and energies because they result from electrons in two different energy levels.

Hybridization of s- and p-orbitals provides sp^3, sp^2 or sp orbitals, each associated with a particular geometry (section 1.1.4). In methane each of four sp^3 hybridized orbitals overlaps with an s-orbital of a hydrogen atom to give a σ-bond. The resulting structure has the four hydrogen atoms arranged tetrahedrally around the central carbon atom (Figure 4).

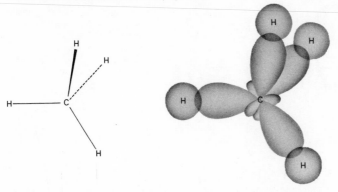

Figure 4 The structure of methane

Ethylene has one π-bond between the two carbon atoms; one p-orbital of each carbon atom is involved in this, and the other three orbitals hybridize to give three sp^2 orbitals trigonally arranged. One of these overlaps with a similar orbital in the second carbon atom, forming the C—C σ-bond; the other two overlap with s-orbitals of two hydrogen atoms to give two C—H σ-bonds. The six atoms in ethylene are therefore coplanar, with the π-bond above and below the plane (Figure 5). Rotation about the C=C bond is energetically unfavourable because any rotation will diminish p-orbital overlap and weaken the π-bond.

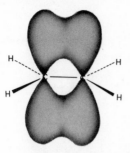

Figure 5 The structure of ethylene

In acetylene, two p-orbitals of each carbon atom are involved in the two π-bonds, and the remaining p-orbital and s-orbital hybridize to give two sp orbitals disposed in diametrically opposite directions. One of these hybrid orbitals overlaps with another of the second carbon atom to give the C—C σ-bond; the other is involved in σ-bonding to the hydrogen atom. The four atoms are therefore collinear.

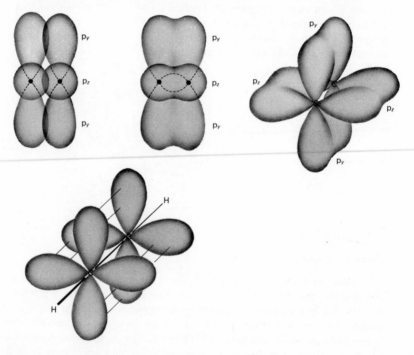

Figure 6 The structure of acetylene

Allene ($CH_2=C=CH_2$) has two separate π-bonds. The central carbon atom (C-2) has two p-orbitals involved in π-bonding and two sp-hybridized orbitals involved in σ-bonding to C-1 and C-3. The two terminal carbon atoms have only one p-orbital used in π-bonding, and so carbon–carbon and carbon–hydrogen bonds employ sp^2-hybridized orbitals. The three carbon atoms are collinear, but the π-electron systems are not because the two p-orbitals at C-2, and so the two derived π-bonds, are at 90° to each other. The structure of allene must therefore be as shown in Figure 7.

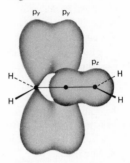

Figure 7 The structure of allene

1.3 Resonance in organic chemistry

Often the properties and reactions of a molecule are not adequately represented by one simple structure. For example, the three carbon–oxygen bonds in the carbonate ion are exactly the same length, although the formal structure makes one oxygen different from the other two (1.1).

$$O=C\underset{O^-}{\overset{O^-}{\diagdown}}$$

(1.1)

Resonance theory attempts to explain such discrepancies. Essentially, all the possible structures which are relevant in explaining the experimental facts are considered. The true structure of the species is then postulated to be something intermediate between these formulae, and more stable than any one. In the case of CO_3^{2-}, three structures can be written

$$O_a=C\underset{O_c^-}{\overset{O_b^-}{\diagdown}} \longleftrightarrow O_a^--C\underset{O_c^-}{\overset{O_b}{\diagdown}} \longleftrightarrow O_a^--C\underset{O_c}{\overset{O_b^-}{\diagdown}}$$

where all the structures are apparently the same, but in which the formal charge resides upon different pairs of oxygen atoms. The true structure of CO_3^{2-} is one in which the carbon–oxygen bond in each case has some double-bond character and the oxygen atoms each have some negative charge (1.2).

$$\left[O = C \underset{\displaystyle O}{\overset{\displaystyle O}{\diagup}} \right]^{2-}$$

(1.2)

The three structures which have been considered are called contributing structures or canonical structures. They do not exist; they are only approximations to show the properties of the true structure. One of the most important warnings must now be given. *The molecule does not spend part of its time as one contributing structure and part of its time as another, nor is the true state of the molecule a mixture of isomers.* It is therefore absurd to talk about a compound 'spending much of its life in the form X'; what is meant is that of all the canonical forms, X most resembles the true structure of the species. This distinction must be clearly understood, for all of the terminology associated with resonance (e.g. 'contributing' structure) is open to this misinterpretation.

The canonical structures are selected on the following basis: (i) each canonical structure must reflect some of the observed properties of the compound; (ii) no canonical structure must involve considerable displacement of the atoms from their normal position (so $[CO_2 + O^{2-}]$ cannot be significant in stabilizing the carbonate ion structure); and (iii) although higher-energy structures may be necessary, excessively high-energy structures cannot be significant since they represent only very minor contributing structures; in other words, the higher the energy, the less the significant contribution it can make.

Resonance theory basically allows for the limitations of our present system of representing structural formulae.* Resonance theory will be often used in this text; we shall now see how it may be applied to the structures of 1,3-butadiene and of benzene.

1.3.1 *1,3-Butadiene*

The formal structure, $CH_2=CHCH=CH_2$, explains the formation of 1,2-dibromo-3-butene when one molecule of bromine adds to one molecule of diene. It does not show why 1,4-dibromo-2-butene is another reaction product. All four carbon atoms in 1,3-butadiene each provide one p-orbital towards π-bonding; the other three orbitals are hybridized to four trigonal sp^2 sets of orbitals. The four carbon atoms and six hydrogen atoms are coplanar, and the four p-orbitals forming the π-bond systems are above and below this plane (Figure 8).

The formal structure of butadiene can be rationalized by considering that overlapping occurs between the p-orbitals of C-1 and C-2, and those of C-3 and C-4. This is artificial; overlapping of the p-orbital on C-2 can occur equally well with that of C-1 or that of C-3, and there is in fact a π-orbital extending over the entire molecule. This explains the resistance to rotation around the C-2—C-3 bond, since the p-orbital overlap must be reduced during twisting. It is therefore possible to talk about two rotational isomers, s-*cis*- and s-*trans*-butadiene. These are

---

* Wheland and Pauling both describe a resonance hybrid by analogy with a mule, which has the properties of an ass and a horse without being either or spending time in either state.

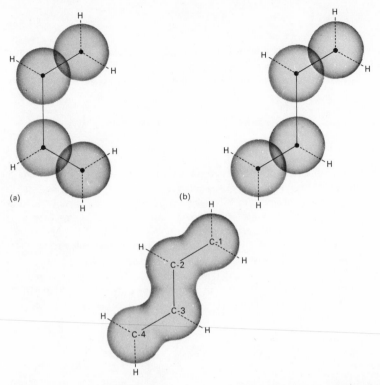

Figure 8 The structure of 1,3-butadiene. (a) s-*cis*-conformer; (b) s-*trans*-conformer

distinguished by the relative positions of the two double-bonds; s-*cis*-butadiene has both ethylenic groups on the same side of the C-2—C-3 bond (Figure 8a), whereas the *trans*-conformer has the ethylenic groups on opposite sides (Figure 8b). It is the result of the double-bond character of the C-2—C-3 bond that rotation of this bond is no longer completely free. Resonance theory explains the 1,4-addition of bromine by considering the structures

$$
\begin{array}{cc}
\begin{array}{l}
\ \ CH_2^- \\
CH \\
\| \\
CH \\
\ \ \ CH_2^+
\end{array}
&
\begin{array}{l}
\ \ CH_2^+ \\
CH \\
\| \\
CH \\
\ \ \ CH_2^-
\end{array}
\end{array}
$$

and

which will also explain the double-bond character of the C-2—C-3 bond and the resistance to rotation around this bond. Notice that the contributing structure $CH_2^+CH^-CH^+CH_2^-$ is not considered, since it is a higher-energy form which has no advantages (in explaining the direction of addition of Br_2) over the less-excited structures, and that

23   Resonance in organic chemistry

$$CH_2=C-CH_2$$
$$\diagdown \diagup$$
$$CH_2$$

is not considered because it involves considerable displacement of the carbon atoms from their normal positions.

The most successful way of writing 1,3-butadiene appears to be

$$CH_2\text{---}CH\text{---}CH\text{---}CH_2,$$

in which each C—C bond is shown to have some double-bond (π-) character.

1.3.2 *Benzene*

No one formal structure can be written for benzene. The Kekulé structures and the Dewar structures localize double bonds between carbon atoms in the ring, yet benzene shows olefinic character only occasionally and usually reacts by substitution rather than by addition. Resonance theory considers all these structures as canonical forms with the true structure of benzene as a hybrid of all of these.

Orbital theory builds up the benzene ring from six carbon atoms, having one p-orbital and three sp^2 orbitals each. Two sp^2 orbitals are involved in carbon-carbon bonding, and the other is used in bonding the substituent (hydrogen etc.). The six p-orbitals, directed above and below the plane of the ring, overlap with either neighbouring orbital equally easily (cf. 1,3-butadiene) and form a closed shell of π-electron density (Figure 9).

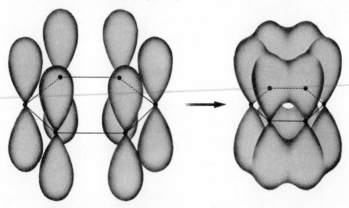

Figure 9 The structure of benzene

Resonance theory demands that such a structure should be more stable than any of its canonical structures. This is apparent from measurements of the heat of hydrogenation of benzene. Cyclohexene liberates 120 kJ mol^{-1} upon hydrogenation; if there is no interaction between double bonds, 1,3-cyclohexadiene ought to show a heat of hydrogenation of 240 kJ mol^{-1}, and 1,3,5-cyclohexatriene (a Kekulé structure of benzene) a value of 360 kJ mol^{-1}. In fact, 1,3-cyclohexadiene has a heat of hydrogenation of 232 kJ mol^{-1}, and benzene a value of 208 kJ mol^{-1}.

Apparently 152 kJ mol^{-1} of energy must be supplied to benzene before it can be considered as 1,3,5-cyclohexatriene. This energy is called the *resonance energy* of benzene and is a measure of the stability of the hybrid relative to one of the canonical structures. Since this stabilization energy results from a closed system of p-orbitals the addition of the first mole of hydrogen (giving 1,3-cyclohexadiene) is endothermic; here most of the stabilization energy is lost. It is also interesting that the cyclic diene seems to have some resonance energy (about 8 kJ mol^{-1}, a similar value to that ascribed to 1,3-butadiene). Not only the fact of resonance stabilization, but also its association with conjugated systems has been demonstrated.

1.3.3 Resonance in phenols

Phenol, C_6H_5OH, shows an acidity in water which is considerably greater than those of the simple alcohols. The dissociation constant (defined for an acid $HX \rightleftharpoons H^+ + X^-$ by the equation $K = [H^+][X^-]/[HX]$) of phenol (10^{-10}) is about a million times greater than that of methanol (10^{-16}) in water at 25 °C; as both ionizations involve C—O—H structures and the heterolysis of an O—H bond, the reason for the difference is not obvious.

The extent to which dissociation of the O—H bond occurs in each case will depend upon the relative stabilities of the acid (ROH) and the anion (RO$^-$), and in particular the extent to which negative charge is localized upon the oxygen atom. In the simple alcohols, there is little chance of the electron being extensively delocalized, but the phenoxide ion allows overlapping between p-orbitals on oxygen and the π-electron system in the ring (Figure 10).

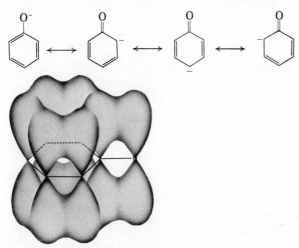

Figure 10 The structure of the phenoxide ion

All four canonical forms will contribute to the resonance structure, which necessarily must be more stable than any of these. As the first structure is the analogue of the alkoxide ion, PhO⁻ is more stable, and hence more readily formed, than RO⁻ and so phenol will be a stronger acid than the alcohols.

1.4 Dipoles in C—X bonding; electronegativity

A covalent bond is formed by the sharing of two electrons through overlapping of individual atomic orbitals, giving a σ-bond or a π-bond. The average positions of these bonding electrons has not yet been considered. Empirically we would expect equal sharing of the electrons only when the two bonded atoms are identical (e.g. H_2); at the other extreme, an ionic bond represents total retention of the 'bonding' electrons by one atom (e.g. Li^+Cl^-). Between these limits there are a number of possibilities in which the electron pair spends more time associated with one atom than with the other. This would be a property of the atoms which were bonded, and would mean that an electron in the vicinity of one atom (e.g. X in the X—Y system) is in a lower-energy state than close by the other atom (Y). Under such conditions, the covalent bond would appear to be polarized; a dipole would be set up between X and Y in which partial negative charge 'resides' at the X end of the bond. The ultimate stage of this unsymmetrical distribution would be the formation of the ions X^- and Y^+.

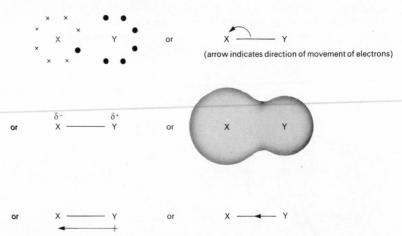

Figure 11 Representations of a dipole in a diatomic molecule

The *electronegativity* of an atom simply describes its tendency to attract electrons to itself, and this in turn is defined by the extent to which the inner shells of electrons shield the nucleus from the valency electrons (which are in the bonding electron shell). A *screening effect* reduces the effective positive charge of the nucleus by the influence of the non-valence (inner) electrons and will be more evident when the valence electrons (the outer shell) are relatively close to the nucleus. If we approximate the electrostatic effects of the positively charged nucleus and all the non-valence electrons as a point charge situated at the nucleus, we can see that in each row of the periodic table, the 'effective nuclear charge' (nuclear charge minus charge of inner-shell electrons) increases as we go from lithium to fluorine. The effect of this upon any valence electron will increase proportionately. Going down any group in the periodic table does not alter the effective nuclear charge, but does increase the distance between the valence electrons and the positive point-source of charge. So the electronegativity of the elements increases across the periodic table (Group I → Group VII) and decreases on going down any one group of the table.

A number of attempts have been made to measure relative electronegativity. The most widely used series is that calculated by Pauling; representative values are given in Table 1. It is worth noting the almost constant increments in any one row of the periodic table.

Table 1 Pauling Electronegativity Values (Relative to Fluorine; F = 4·0)

H	2·1												
Li	1·0	Be	1·5	B	2·0	C	2·5	N	3·0	O	3·5	F	4·0
Na	0·9	Mg	1·2	Al	1·5	Si	1·8	P	2·1	S	2·5	Cl	3·0
K	0·8	Ca	1·0	Sc	1·3	Ge	1·7	As	2·0	Se	2·4	Br	2·8
Rb	0·8	Sr	1·0	Y	1·3	Sn	1·7	Sb	1·8	Te	2·1	I	2·4
Cs	0·7	Ba	0·9										

Elements with greatly different electronegativities form ionic bonds; those with more similar electronegativities (e.g. C—Halogen, B—H) form covalent bonds whose direction and intensity of polarization can be deduced from the Pauling values.

The fact that many covalent bonds are dipolar is invaluable in explaining the behaviour of organic molecules.

.5   Inductive, mesomeric and field effects

Two dissimilar atoms, bonded together, set up a dipole between them. The extents of these dipoles have considerable bearing upon the courses of many reactions. For instance, since acidity (in the Brønsted–Lowry sense) involves the formation of H^+, we could expect that the O—H bond is more polar, and hence more likely to form H^+ by complete heterolysis, than the C—H bonds (comparing the electronegativities of the elements involved). Methanol is therefore more acidic than methane because of its O—H bond, and the relative ease with which this bond heterolyses.

The *extent* of this dipole may be altered by other groups in the system, and in organic chemistry we find such substituents having a remarkable effect. To discuss substituent effects we need two fundamentals: (i) a standard state, and (ii) mechanisms for transmitting the effect to the reaction site. The standard state is taken as the behaviour of the fully hydrogenated (parent) system. In short, *hydrogen is the standard substituent.*

We can distinguish three mechanisms by which substituent effects may be transmitted to the reaction centre electrically. Either the π-electrons may be involved, in which case there is conjugative (or mesomeric) relay, or the σ-electrons are perturbed (giving rise to inductive relay), or the effect may be transmitted across empty space (field effect).

1.5.1 *Field effect*

The field effect is due to electrostatic interaction between a charged or partially charged atom and a reaction centre. If, for instance, a positively charged group (e.g. $-NMe_3^+$) was in the vicinity of a reaction centre (as in o-$NMe_3^+.C_6H_4.CO_2Me$) there would be considerable resistance to that reaction centre acquiring a positive charge. If the reaction proceeded through such a positively charged intermediate, then the reaction itself would be impeded. Conversely, if the reaction proceeded with the formation of negative charge at the reaction site (i.e. $C^{\delta-}$) it would be facilitated by the electrostatic field.

Such a field effect is difficult to detect unambiguously. Since it involves electrostatic interaction, it must only be readily detectable with a group very close to the reaction centre. Under these conditions, the other electronic effects are at their greatest and, in addition, there is the possibility of steric effects confusing the picture.

1.5.2 *Inductive effect*

This effect is well exemplified in the acidity of acetic acid, which is vastly increased by successive substitution of hydrogen by chlorine. Monochloroacetic acid, for instance, is about sixty times the strength (from dissociation-constant measurements) of acetic acid. The reason is the marked C—Cl dipole.

The C—Cl dipole induces charge upon the adjacent atoms, and the effect is then relayed throughout the system. This is the *inductive effect*, which operates mainly through the σ-bonds.

$$Cl \leftarrow \underset{\delta+}{C} - C - O - H \quad \text{causes}$$
$$\underset{\|}{O}$$

$$Cl \leftarrow \overset{\delta\delta+}{C} \leftarrow C - O - H \quad \text{causes}$$
$$\underset{\|}{O}$$

$$Cl \leftarrow C \leftarrow \overset{\delta\delta\delta+}{C} \leftarrow O - H \quad \text{causes}$$
$$\underset{\|}{O}$$

$$Cl \leftarrow C \leftarrow C \leftarrow \overset{\delta\delta\delta\delta+}{O} - H \quad \text{causes higher acidity}$$
$$\underset{O}{\|} \qquad \text{of the hydrogen atom.}$$

Two chlorine atoms substituted at the α carbon atom will each exert an electron-withdrawing effect and will increase the electronegativity of this carbon atom, and hence the magnitude of the effect relayed to the O—H bond. As we have seen, the acidity of acetic acid increases as successive hydrogen atoms are replaced by chlorine.

Since this effect is relayed by induction, its magnitude should decrease as the distance (number of atoms) between the dipole and the reacting group increases. Although increasing the length of the alkyl chain attached to the $-CO_2H$ fragment does not materially alter the dissociation constant ($K = 1\cdot4 \times 10^{-5}$ for pentanoic acid), the dissociation constants of the ω-chloroacids drop regularly:

$$ClCH_2CO_2H \quad K = 155 \times 10^{-5},$$
$$ClCH_2CH_2CO_2H \quad K = 8\cdot2 \times 10^{-5},$$
$$ClCH_2CH_2CH_2CO_2H \quad K = 3\cdot0 \times 10^{-5},$$
$$ClCH_2CH_2CH_2CH_2CO_2H \quad K = 1\cdot9 \times 10^{-5}.$$

The inductive effect therefore dies out fairly quickly as successive atoms are interspersed between the dipole and the reaction centre.

1.5.3 Mesomeric effect

We have already seen that in the case of 1,3-butadiene two apparently isolated double bonds can act as one continuous π-orbital (section 1.3.1), and that the three formal double bonds in benzene are, in fact, one molecular π-orbital (1.3.2). If a group is attached to such a conjugated system, and if p-orbital overlap can occur between the attached atom (X) and the π-system in the conjugated structure, electron density within the conjugated system may alter. If the group, X, repels

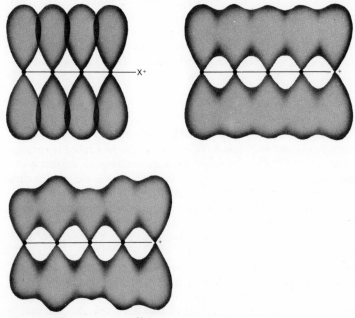

Figure 12 The mesomeric effect

electrons from the overlapping p-orbital into the π-orbital system, increases in electron density will occur at alternate carbon atoms in the conjugated structure. Unlike the inductive effect, this *mesomeric effect* is transmitted through π-bonds with very little loss of intensity. It is greatest when the extent of pπ-orbital overlap is greatest; since carbon is a relatively small atom, the elements of the first row of the periodic table show stronger mesomeric effects than the larger atoms further down the groups (i.e. $F > Cl > Br > I$).

Resonance theory considers contributions from structures such as

$$C=C-C=C-X \quad \longleftrightarrow \quad \begin{array}{c} \bar{C}-C=C-C=X^+ \\ \text{and} \\ C=C-\bar{C}-C=X^+ \end{array}$$

with the same overall effect that charge appears upon alternate atoms of the conjugated system.

1.5.4 *Inductive and mesomeric effects in organic chemistry*

Taking hydrogen as the standard, an electron-attracting substituent (X) is said to show a −I effect (if it operates inductively) or a −M effect (if mesomeric effects are involved in this electron attraction). The halogens are therefore said to show a −I, +M effect; electrons are withdrawn by the inductive effect, but donated by the mesomeric effect. The same is also true of oxygen and nitrogen substituents like −OR, −OH, −NH$_2$, −NMe$_2$, but not −NO$_2$ or −NMe$_3^+$. If the substituent attracts electrons less strongly than hydrogen, it is said to show a +I effect (inductively) or a +M effect (mesomerically). Alkyl groups show a weak +I effect (cf. the dissociation constants of CH_3CO_2H (1.8×10^{-5}) and HCO_2H (1.7×10^{-4})). The nitro group shows a strong −M effect, due to contributions of structures such as (1.3).

$$\underset{+}{C}-\underset{+}{C}=N\overset{O^-}{\underset{O^-}{\diagdown}} \cdot$$

(1.3)

The unusually acidic properties of phenol have already been described (section 1.3.3) and the unusually weak basicity of aniline can also be ascribed to conjugative (mesomeric) effects. The lone pair of electrons on nitrogen is situated in an orbital parallel with the π-electron system of the benzene ring, and these electrons can therefore be delocalized into the benzene ring system. However, it is by coordination of these electrons with H^+ that aniline shows its basicity, and since the effect of mesomerism is to diminish the electron density on nitrogen, the basicity (compared with NH_3 or CH_3NH_2) is decreased (by a factor of 10^6 in water at 25 °C).

This carries with it the implication that the benzene ring in aniline shows an increased electron density (relative to PhH itself) at the *ortho* and *para* positions; this is in fact detectable (Chapter 2). Notice that the M effect is assumed to outweigh the I effect of nitrogen; in almost all cases this is true, although with

strongly electronegative groups such as the halogens the different intensities of
the two effects becomes comparable.

These two electronic effects are thought to be present in the ground state of
the molecule. Other, similar effects may facilitate the course of the reaction; in
other words, further I or M effects may become apparent to meet the needs of
the reagent. The extent of these polarizability effects is, of course, dependent
upon the demands of the reaction. The measurement of these effects and even
their unambiguous detection is difficult.

5.5 *The acidities of phenols and of carboxylic acids*

The acidity of phenol itself has been explained by the resonance stabilization of
PhO^-. The extent to which PhO^- is stabilized could be enhanced or diminished
by a substituent in an *ortho* or *para* position. An $-NO_2$ group, for instance, allows
the extra contributing structure (1.4) which will increase the overall stability of

(1.4)

the *p*-nitrophenoxide ion and so cause *p*-nitrophenol to be more acidic (by a
factor of about 10^3 in water at 25 °C) than phenol itself. Further substitution of
$-NO_2$ in the *ortho* positions enhances the stability of the anion, and hence the
acidity of the phenol, even further until ultimately picric acid ($K = 0.88$) is
obtained. Nitro groups in the *meta* position, where such stabilization is not
possible (except by a higher-energy structure) have a much less dramatic effect.

Similarly, +M groups in the *ortho* or *para* positions (e.g. *p*-OMe.C_6H_4.OH,
$K = 6 \times 10^{-11}$; *p*-$^-O.C_6H_4$.OH, $K = 9 \times 10^{-13}$) decrease the acidity of phenol.

In the case of benzoic acid, the carboxylate anion is already resonance
stabilized (1.5) but suitable conjugation between groups in the benzene ring

(1.5)

and the carboxylate fragment does not occur. However, the conjugative effect
can be transmitted from a group in the *ortho* or *para* position to carbonyl carbon,
from which it must be transmitted to oxygen or the O—H bond by an inductive
effect. Any resonance effect from such a group would therefore be felt only
slightly, since this is a second-order effect (i.e. indirect). Accordingly, the $-NO_2$
group in the *para* position of benzoic acid increases the dissociation constant by
a factor of six (cf. 1000 for the same situation in phenol), while —OH and —OMe
groups similarly situated halve the dissociation constant.

Problems

1.1 One of the fundamental objections to an atomic theory involving satellite electrons spinning around the nucleus was that the electron ought to lose energy in radiation and hence inevitably fall into the nucleus. Does the problem still exist with the present theory?

1.2 In which directions are the following bonds polarized: Br—F, Br—C, Br—N, C—N, N—O?

1.3 Although most hydrocarbons are feebly acidic ($K \simeq 10^{-40}$), some C—H bonds are remarkably more acidic. Examples are 1,3-cyclopentadiene, diethyl malonate and nitromethane. Give reasons for these observations.

1.4 Which of the following pairs of p-substituents will have the greater effect upon the acidity of phenol, and in which direction?

(a) —NH_2 or —PH_2 (d) —NO_2 or —NHOH

(b) —NH_2 or —CH_3 (e) —OMe or —SMe

(c) —NH_2 or —NH_3^+ (f) —F or —Br.

1.5 How would a p-NO_2 group affect the basicity of aniline?

1.6 Does the spatial requirements of sp, sp^2, and sp^3 hybridized carbon atoms suggest any particular direction of attack in additions to olefins and acetylenes?

1.7 Draw the contributing resonance structure of aniline showing that the electron density of the nitrogen atom is decreased.

1.8 Classical concepts imply that the inductive effect may only be transmitted through σ-bonds, while mesomeric effects can only be relayed through an unbroken π-system. Is this limitation reasonable?

Chapter 2
General Mechanistic Considerations

Classification of reactions and of reagents

Chapter 1 dealt with the formation of interatomic bonds between atoms, which could involve one electron being transferred (ionic bonding), or two electrons being shared; either each atom provides one of the bonding electrons (covalent bonding) or both electrons are denoted by one of the bonded atoms (dative bonding). Bond-breaking processes will now be discussed.

1.1   *Heterolysis and homolysis*

The bond in the molecule $A:B$ may be broken in two ways. If each fragment retains one of the bonding electrons,

$$A:B \; \rightarrow \; A \cdot + B \cdot, \qquad\qquad\qquad 2.1$$

a *homolytic* fission is said to occur. The covalent bond can also be broken so that both bonding electrons are retained by one of the fragments,

$$A:B \rightarrow A + B:. \qquad\qquad\qquad 2.2$$

This process is called *heterolysis.*

Although only two of the valence electrons are shown, the others are implied in these formulae. The products of a homolytic fission are free *radicals* (species with an unpaired electron in the outer shell), while the products of heterolytic fission are *ions*. The products of equation 2.2 may be written A^{+} and B^{-}, shown in the following examples,

$$\underset{\times\times}{\overset{\times\times}{\times}}\ddot{C}l \times \ddot{C}l: \quad \xrightarrow{\text{homolysis}} \quad \underset{\times\times}{\overset{\times\times}{\times}}\ddot{C}l^{\times} \quad \cdot\ddot{C}l:$$

$$\underset{\times\times}{\overset{\times\times}{\times}}\ddot{C}l \times \ddot{C}l: \quad \xrightarrow{\text{heterolysis}} \quad \underset{\times\times}{\overset{\times\times}{\times}}\ddot{C}l \quad \times\ddot{C}l:$$

in which only the outer electrons are shown.

Two types of reaction are now distinguishable, depending upon whether the process involves free radicals (homolytic process) or whether it involves ionic species (heterolytic or ionic process). It is the nature of the bond-breaking reactions, and not the nature or charge of the attacking species, which defines the mechanism.

2.1.2 *Electrophiles and nucleophiles*

Within the heterolytic processes there are two further distinctions. If the process is initiated by a species which attacks sites of relatively high electron density, it is said to be an *electrophilic* process, and the species which attacks is said to be an electrophile. If the process is initiated by a species seeking to donate electrons to sites of relatively low electron density, a *nucleophilic* reaction takes place, and the species is called a nucleophile. The formal charge upon the attacking species is not a guarantee of its mode of reaction.

In the reactions

$$H_3O^+ + OH^- \rightleftharpoons 2H_2O,$$
$$H_3O^+ + NH_3 \rightleftharpoons NH_4^+ + H_2O,$$
$$H_3O^+ + H_2NCH_2CH_2NH_3^+ \rightleftharpoons H_2O + H_3N^+CH_2CH_2NH_3^+,$$

the three bases are all nucleophiles towards H_3O^+, *regardless of their formal charge*; a Brønsted–Lowry base is simply a substance which is nucleophilic towards protons. Not all compounds which are nucleophilic towards protons (and are therefore bases) are nucleophilic towards carbon. For example, 2,6-dimethylaniline shows basic properties but only reacts very slowly with CH_3I, although ammonia readily undergoes nucleophilic attack by both H_3O^+ and CH_3I. Conversely, very little reaction occurs between H_3O^+ and I^- ($K \simeq 10^{-10}$), although iodide ion is quite a strong nucleophile towards carbon. This is true if we consider nucleophilic power in terms of the position of an equilibrium; however, it is not true if we consider that the rate of reaction defines nucleophilicity, since the neutralization reaction occurs much more rapidly, although to a lesser extent (less-favourable equilibrium position).

In discussing organic reactions, the organic molecule is regarded as the substrate (the compound being acted upon) and the other species as the reagent. Hence the reaction

$$I^- + CH_3Cl \rightarrow CH_3I + Cl^-$$

is a nucleophilic process, since iodide ion attacks the carbon atom in methyl iodide and gives electrons in forming the C—I bond. The reaction

$$H_3O^+ + H_2C=CH_2 \rightarrow [CH_3CH_2]^+ \xrightarrow{H_2O} CH_3CH_2OH$$

is an electrophilic process, for it is initiated by proton attack upon the π-bond of ethylene. The essential distinction is that the reagent may form a bond either by donating electrons (nucleophilic attack) or by accepting them (electrophilic attack).

2.2 Reaction profiles

A reaction profile is an attempt to show the course of a reaction in more detail than that provided by a simple chemical equation. Suppose that we are interested in the process

$$OMe^- + EtCl \rightarrow EtOMe + Cl^-$$

and we wish to know *how* this displacement takes place. It seems that OMe⁻ has
displaced Cl⁻, and presumably by attack at the positive end of the $C^{\delta+} - Cl^{\delta-}$
dipole. How can the details of the reaction be described?

Imagine the two reagents separated so that they do not interact with each other
at all (i.e. in the ground state) and consider the effect of bringing the reactants
together. Initially, as OMe⁻ approaches the alkyl halide (presumably in the same
line as the C—Cl bond but from the carbon side, since this would involve the least
repulsion by the similarly charged chlorine atom), its electric field would increase
the C—Cl bond length, and the degree of polarity of the C—Cl bond. Both
stretching and polarizing the bond would require energy to be put *into* the system:

$$MeO^- \qquad\qquad C{-}Cl$$
$$MeO^- \text{-------} \overset{\delta+}{C}{-}\overset{\delta-}{Cl}$$
$$MeO^- \text{----} C \text{---} Cl$$

$$\overset{\longleftrightarrow}{\underbrace{\qquad\qquad\qquad}}$$
(energy put in)

At some point, however, a carbon–oxygen bond begins to be formed and this
involves some energy being *released.*

As the reagent (OMe⁻) approaches still further, the extent of carbon–oxygen
bond formation increases rapidly until the energy gained from this process outweighs the rather small energy needed to complete the breaking of the C—Cl bond
and the formation of Cl⁻. From this stage energy is given out until the products are
in their ground states: that is, when Cl⁻ and EtOMe are fully formed and separated
by so great a distance that there is no interaction between them.

$$MeO{-\!-\!-\!-}C \text{------} \overset{\delta-}{Cl}$$
$$MeO{-}C \text{--------} Cl^-$$
$$MeO.C \qquad\qquad Cl^-$$

$$\overset{\longleftrightarrow}{\underbrace{\qquad\qquad\qquad}}$$
(energy given out)

All this information, more digestibly presented, is contained in the reaction
profile (Figure 13) which plots the energy of the system against the reaction
coordinate (a measure of the extent to which the reaction has taken place).
Reading from the left, the initial, endothermic process involves bringing the
reagents together to the point where they react together spontaneously (exo-
thermically). As the reagent approaches, the C—Cl bond would become more
polarized (by induction) and longer; similarly, the attraction between OMe⁻ and
$C^{\delta+}$ would increase and some degree of bonding between the two could occur.
Polarizing the C—Cl bond and lengthening it would require energy, and so the total
energy of the process will increase until some stage when the energy of C—O bond
formation outweighs the steadily decreasing C—Cl bond energy. At this stage the
process is exothermic until the products are formed in their ground states.

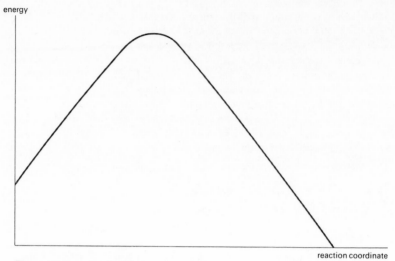

Figure 13 Reaction profile

$$\text{MeO}^- \qquad \overset{\delta+\ \delta-}{\text{C—Cl}} \qquad \text{(reagents: ground state)}$$

$$\text{MeO}^- \text{------} \text{C}^+ \text{----} \text{Cl}^-$$

$$\text{MeO}^- \text{--} \text{C}^+ \text{----} \text{Cl}^-$$

$$\text{MeO—C} \qquad\qquad\qquad \text{Cl}^- \text{(products)}$$

The point at which the system has the highest energy is the *transition state*. At this particular configuration it is equally likely that the reactants can give products or return to starting materials. The point at which the energy of the system reaches a maximum is decided by the bond strengths of the C—O and C—Cl bonds, the electronegativities of the two atoms (O and Cl), and the heats of formation and of solvation of the halide ion, of the methoxide ion and of the dipolar intermediates. For these reasons the position of the transition state along the reaction coordinate will differ in different chemical reactions.

The energy needed to bring the starting materials from the ground state to the transition state is the *activation energy of the forward reaction* (E_1, Figure 14); the products are separated from the transition state by the *activation energy of the reverse reaction* (E_{-1}). The overall *heat of reaction* therefore is the difference between these two energies, $H = E_1 - E_{-1}$.

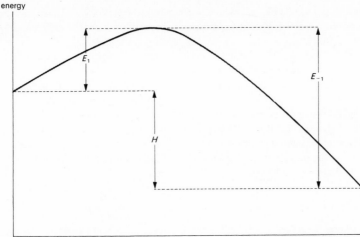

Figure 14 Reaction profile showing energy changes

2.3 Multi-stage reactions

The reaction of methoxide ion with methyl chloride is a simple process, involving only one transition state. In contrast, the base-catalysed hydrolysis of chloroform

$$HCCl_3 + OH^- + H_2O \rightarrow HCO_2^-, CO, \text{ and } Cl^-$$

has been shown to involve several stages.

Initially a proton is removed from chloroform to give the trichloromethyl anion CCl_3^-. This is a rapidly established equilibrium which precedes the slow stage of the

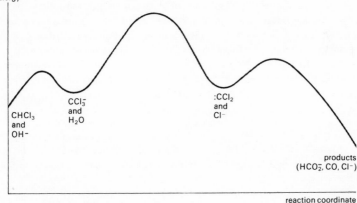

Figure 15 Reaction profile of the base-catalysed hydrolysis of chloroform

sequence, the loss of Cl^- to give $:CCl_2$. Dichlorocarbene then rapidly reacts with the solvent to provide (possibly through a number of intermediate compounds) the reaction products.

In this sequence, CCl_3^- and $:CCl_2$ are *intermediates*. They have some stability with respect to their immediate environment (in other words, they are *not* the highest-energy configurations of the system) and they are not transition-state complexes. In the reaction, however, they pass through transition-state configurations, a process which again requires activation energy.

The differing degrees of stability of intermediates means that their presence may be definitely shown or merely inferred. A transition-state complex, however, cannot be shown to exist because it is energetically unstable with respect to its surroundings.

2.4   Solvent effects

Since heterolysis involves the formation of ions (a process involving high energy because of the separation of unlike charges) it is much more sensitive to the effects of solvents than homolysis. Since solvation energies may reach 420 kJ mol^{-1} in a polar solvent (e.g. water, methanol) there will be many reactions in which a polar solvent is essential to make them thermodynamically possible, and so small changes in solvent polarity may have considerable effects upon the rates and extents of such processes.

In general, polar solvents (those which can dissolve salts to some extent* and which therefore stabilize ionic species in solution) are dipolar molecules with electron-donating properties. Usually an atom such as oxygen or nitrogen (which have lone pairs of electrons) is associated in such structures. Amides, nitriles, nitro compounds and alcohols are fairly good polar solvents; ketones and sulphoxides (especially Me_2SO, dimethyl sulphoxide) are also used.

The polarity of a solvent might seem to be directly linked to its dielectric constant. In fact this seems to be true only within a series of similar compounds (e.g. alcohols), and the polarity of solvents of similar dielectric constant (e.g. MeOH, $HCONMe_2$ and $MeNO_2$) is certainly not the same. The absolute measure of solvent polarity has not been possible; a number of parameters have been measured which reflect polarity in certain reactions, but none of these is completely suitable.

2.4.1   *Hughes–Ingold theory of solvation*

Without an absolute measure of solvent polarity, a number of predictions can be made from the demands of the different chemical reactions. The Hughes–Ingold theory considers the structure of the reagents, and compares it with that of the transition state.

---

* In dissolving an ionic crystal, work must be done in breaking down the crystal and in stabilizing the liberated ions. If a solvent can achieve this, it is also probable that it will assist in the heterolysis of a covalent bond.

When electrical charge is created or localized in a reaction intermediate, the rate of the reaction is increased by increasing the polarity of the solvent, and decreased by decreasing solvent polarity. For example, the rate of formation of $NMe_4^+I^-$ from MeI and NMe_3 would be accelerated by more polar solvents (e.g. EtOH) and retarded by non-polar solvents (e.g. PhH). Conversely, when charge is dispersed or destroyed during the reaction, the process will be retarded by polar solvents and accelerated by non-polar solvents. The reverse reaction,

$$NMe_4^+I^- \rightarrow NMe_3 + MeI,$$

would be an example of such a process. Depending upon the relative degree of charge dispersal or concentration, the given effect of any change of solvent properties would be large or small; in a situation in which charge is only slightly dispersed in the transition state, only slight changes in rate would be expected.

2.4.2   Grunwald–Winstein Y-values

The Grunwald–Winstein measure of solvent polarity makes use of a particular mode of solvolysis of organic halides. In a good polar solvent, t-BuCl reacts by a mechanism in which the rate-determining stage involves only the formation of a carbonium ion or a similar structure:

$$t\text{-BuCl} \xrightarrow{\text{slow}} t\text{-Bu}^+$$
$$t\text{-Bu}^+ + X^- \xrightarrow{\text{fast}} t\text{-BuX} \quad (X = \text{any nucleophile}). \hspace{2cm} 2.3$$

This is known as the S_N1 mechanism (see Chapter 4).

Since the rate of the reaction depends upon the *formation* of ions, it follows that variations in the reaction rate within different solvents should parallel the polarity of the solvent. Because the second stage is fast, the nature of the nucleophile cannot affect the overall reaction rate. The rate of the reaction of t-BuCl with a nucleophile was measured in 80 per cent ethanol (taken as standard) and in a solvent of unknown solvent polarity, both at 25 °C. Taking a parameter Y, which measures solvent polarity, and allowing $Y = 0$ for 80 per cent ethanol,

$$\log \frac{k}{k_0} = Y,$$

where k and k_0 are the rates in the unknown solvent and in 80 per cent ethanol respectively. For other halides, reacting by a similar mechanism, the equation becomes

$$\log \frac{k}{k_0} = mY,$$

where m measures the sensitivity of the organic halide to solvent effects, relative to t-BuCl ($m = 1$). In this way a number of Y-values were measured for different solvents and solvent mixtures. The treatment was not totally successful, for m was found to vary with the natures of the solvents used (i.e. m for Ph_2CHBr differed depending upon whether aqueous acetone, aqueous acetic acid, or aqueous dioxan was used); the equation does not have the general usefulness which was anticipated. It is also restricted to solvents which were sufficiently polar to allow the S_N1 reaction to occur; since another mechanism becomes preferred at lower solvent polarities, and since a solvolysis reaction (X^- = solvent) shows the same kinetic behaviour for either mechanism, the Grunwald–Winstein equation had real significance only in those solvents in which the S_N1 mechanism could be unambiguously proved.

2.4.3   *Swain equation*

Swain considered all ionic displacement reactions to involve three molecules in the rate-determining stage. For the reactions of an alkyl halide RCl,

Rate = k[X] [RCl] [Y].

One molecule (X, say) supplied a nucleophilic push to the RCl molecule, and the other (Y) supplied an electrophilic pull to the departing Cl^-:

$$\overset{\curvearrowright}{X} \quad \overset{\delta+}{R} \cdots \overset{\delta-}{Cl} \quad \overset{\curvearrowright}{Y}.$$

Either X or Y or both could be solvent molecules; since the two possible proposed mechanisms of reaction of organic halides were replaced by the single, termolecular mechanism, and heterolytic reaction of RCl would serve to define a solvent parameter. Relative rates of reaction were defined in terms of two sets of parameters. One set reflected the sensitivity of the reaction to electrophilic 'pull' (s_e) and to nucleophilic 'push' (s_n); the other set measured the reactivities of X and Y as nucleophiles (n) and electrophiles (e) respectively:

$$\log \frac{k}{k_0} = es_e + ns_n. \qquad\qquad 2.4$$

Again, *t*-BuCl was selected as standard ($s_e = s_n = 0$) and 80 per cent ethanol was taken as the reference solvent ($n = e = 0$). Two other ratios were necessary to define the limits of the equation. For MeBr, s_n/s_e was defined as 3·00; for Ph_3CF, s_e/s_n was also defined as 3·00.

The resulting values of the various parameters gave the separate contributions of any solvent as an electrophile and as a nucleophile. The values paralleled chemical experience in general; however, there are sufficient inconsistencies (even if the fundamental premise of a termolecular mechanism is allowed) to restrict the use of the equation.

2.4.4   *Kosower Z-values*

The ultraviolet spectrum of a pyridinium salt (e.g. 2.1) shows a strong band due to a charge-transfer complex resulting between donor and acceptor molecules in which an electron appears to be transferred, giving unusual spectroscopic properties. An example is the hydrocarbon–picric acid complex, or the species formed reversibly when benzene and iodine solutions are mixed.

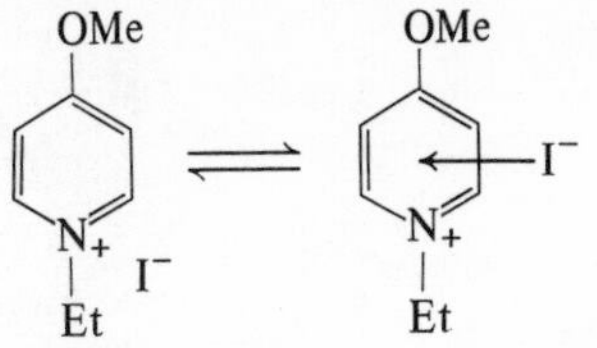

OMe   OMe

$\rightleftharpoons$

N+   N+
| I⁻ | ⟵—I⁻
Et   Et

(2.1)

The position of this band in the u.v. spectrum depends upon the polarity of the solvent in which the salt is studied; the position varies linearly with Grunwald–Winstein Y-values for a number of aqueous solvents. A Z-value, corresponding to the u.v. position of maximum absorption of this charge-transfer complex,* is then defined. It has the advantage of being measurable in non-nucleophilic solvents and therefore extends the range of the Grunwald–Winstein treatment. In extending the range, one must assume that the difference in the nature of the process being studied still does not cause difficulties in interpretation. However, the factors involved in exciting an electron to a higher state need not be the same, nor proportional, to those involved in forming a transition state. Comparing spectral and kinetic results need not therefore be valid.

5 Salt effects

A reaction in which ions are formed, or where charge is created or concentrated, will be subject to a considerable solvent effect. Dissolved ionic substances in the reaction medium will also have a detectable effect; depending upon the nature of the ion and of the reaction, these salt effects may be discussed under three main headings.

5.1 Positive primary salt effects

A primary effect has a direct influence upon the reaction under study. In the hypothetical process,

$$A : B \xrightarrow{\text{slow}} A^+ + B^-$$

charge is formed or created, and so any factor which will stabilize ionic charge will accelerate the reaction, or at least make it energetically more likely. One of these factors is a polar solvent. Since an ion (X^- in Figure 16) could replace one of the solvent molecules around either A^+ or B^-, and by doing so would considerably stabilize it, an added salt would accelerate the heterolysis reaction. This simple picture suggests that the extent of acceleration would depend upon (i) the concentration of salt in solution and (ii) the charge upon the ions constituting the salt. This is in general agreement with experiment; LiCl is less effective than

* Z-values correspond to the energy of the transition ($E = h\nu$).

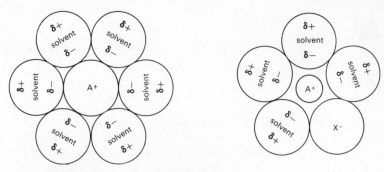

Figure 16 'Solvating' effect of the X⁻ anion

MgCl$_2$ at the same molar concentration, but increasing concentrations of either salt cause greater acceleration of the rate. This effect, shown by all salts, is a *positive* (i.e. accelerating) primary salt effect.

2.5.2   *Negative salt effects*

Consider the reaction sequence

$$A:B \xrightarrow{\text{slow}} A^+ + B^-; A^+ \xrightarrow{\text{fast}} \text{products.}$$

(Compare this with equation 2.3, and the S$_N$1 mechanism in Chapter 4.) Because the rate-determining (slow) stage of the reaction involves heterolysis, the rate of the reaction will be accelerated by polar solvents and it will show a positive primary salt effect. However, if salts containing the B⁻ anion are introduced into the solution the position of the equilibrium between AB, A⁺ and B⁻ is disturbed and the rate of reaction (i.e. the rate of formation of products) will be diminished, although the salts presumably show a positive salt effect as well. Since this 'common-ion' effect retards the rate of the reaction, it is called a *negative primary salt effect*. The usual result of these two opposing effects is a slight retardation in rate; occasionally the common-ion effect appears as an unusually small positive salt effect.

   Another effect is found with the conjugate base of the reaction solvent. Ions such as OEt⁻ (in EtOH) and OH⁻ (in aqueous media) show a specific and slight retarding effect upon some solvolysis reactions. The cause of this is not fully understood; it is a negative effect, but whether primary or secondary is uncertain.

2.5.3   *Secondary salt effects*

A salt can also have an indirect effect upon a reaction. In the sequence

$$HX \underset{}{\overset{\text{fast}}{\rightleftharpoons}} H^+ + X^-,$$

$$H^+ + A \xrightarrow{\text{slow}} \text{products,}$$

the rate of formation of products will depend upon the concentrations of both A and H⁺ (i.e. acid-catalysed). While salts will generally show a positive *secondary* salt effect (since they promote the ionization of the acid HX), a *negative secondary* effect will be found with salts having the X⁻ anion, for these will repress the ionization of HX and hence, by lowering the $[H^+]$ in solution, lower the rate of formation of products.

.6   Isotope effects

Substituting deuterium for hydrogen in the C—H bond increases the bond strength between the two atoms. In a reaction where there is partial or total breaking of the C—H bond, these differences in energy will reflect themselves in the rates of the two reactions. An example is the rate of substitution by iodine of 2,4,6-trideuterophenol (2.2), which proceeds at one quarter the rate of ordinary phenol.

(2.2) D

Although these differences must be apparent in the ground state and the transition state, the magnitude of the observed effect implies that often the whole potential isotope effect is used in passing from the reagents to the transition state. If the rest of the molecule is 'heavy' compared with hydrogen, the ratio of rates k_H/k_D is dependent upon $(m_D/m_H)^{\frac{1}{2}} = 1 \cdot 414$. Since this ratio rapidly diminished with heavier elements ($^{12}C/^{14}C = 1 \cdot 08$; $^{35}Cl/^{37}Cl = 1 \cdot 03$) hydrogen isotopes are the most easy to detect unequivocally because of their size. The assumption is usually made that isotopes do not show different substituent effects, so that any differences in reactivity between isotopically different systems must be due to an influence upon a bond-breaking process in the rate-determining step of the reaction sequence.

.6.1   *Solvent isotope effects*

A reaction which involves exchange between the reagents and the solvent during or prior to the rate-determining stage will show some effect if the reaction is studied in D_2O and H_2O. An example of this is the base-catalysed (NaOEt) formation of styrene from β-bromoethylbenzene

$$PhCH_2CH_2Br \xrightarrow[EtOH]{NaOEt} PhCH=CH_2. \qquad\qquad 2.5$$

One proposed mechanism of the reaction involved the anion $Ph\bar{C}HCH_2Br$, which was formed by the rapid equilibrium

$$PhCH_2CH_2Br \underset{EtOH}{\overset{OEt^-}{\rightleftarrows}} Ph\bar{C}HCH_2Br.$$

If this mechanism were true, the reaction would give $PhCHDCH_2Br$ and $PhCD_2CH_2Br$ when carried out in EtOD. In fact, unreacted β-bromoethylbenzene extracted from such a reaction mixture showed no uptake of deuterium, disproving the proposal.

Since D_2O is a weaker base than H_2O, solvent isotope effects can be shown in a reaction such as

$$A + D_3O^+ \xrightleftharpoons{\text{fast}} AD^+ + D_2O, \qquad\qquad 2.6$$

$$AD^+ \xrightarrow{\text{slow}} \text{products.} \qquad\qquad 2.7$$

The position of the pre-equilibrium (equation 2.6) will be more to the left in H_2O than in D_2O, because of the differing acidities of H_3O^+ and D_3O^+. At stoichiometrically identical concentrations of acid and of A (i.e. the same concentration of acid HX used to provide H_3O^+ or D_3O^+) more B^+ will exist in the D_2O solution. Since the reaction rate depends upon [B], the overall rate of reaction will be faster in the deuterated solvent.

Such a situation is found in the chlorination of aromatic compounds by acidified solutions of HOCl. Again it is usually assumed that the isotopically different solvents have identical solvation properties, so that any differences in reactivity are attributed to the effect of the isotope upon the energy of the transition state.

2.7 Catalysis by acids and by bases

Acid catalysis may either involve Brønsted–Lowry acids (those which supply H^+, e.g. HCl, H_3O^+) or involve Lewis acids (those which can accept a pair of electrons from a base to form a coordinate bond, e.g. $BF_3 \rightarrow BF_4^-$, or $H^+ \rightarrow H_3O^+$). Lewis acids are used as catalysts in a number of organic processes (e.g. Friedel–Crafts acylation and alkylation, Gattermann–Koch reaction and catalysed halogenation), but it is more convenient to deal with these reactions individually. In the present section we shall confine our attention to catalysis involving Brønsted acids and their conjugate bases.

2.7.1 *Acid catalysis*

In the reaction

A → products,

two possible types of acid catalysis can be described. *Specific acid catalysis* takes place when a reaction is accelerated only by hydrogen ion, so that the kinetic equation becomes

$$\text{Rate} = k[H^+][A] \qquad\qquad 2.8$$

in the process

$$A + \text{acid} \xrightarrow{\text{slow}} \text{products.}$$

Since H^+ is a very efficient catalyst in aqueous media it is often experimentally difficult to detect the alternative *general acid catalysis*. This occurs when all the acids in solution (not only H^+) catalyse the reaction. The kinetic equation then becomes

$$\text{Rate} = k[H^+][A] + k'[HX][A] + k''[A], \qquad\qquad 2.9$$

reflecting the contributions by H^+, by the un-ionized acid HX (e.g. acetic acid) in solution, and by H_2O (k'' should strictly be rendered k''' $[H_2O]$). The last two terms may be distinguished by a number of studies in buffer solutions, but because the first term is usually the largest, general acid catalysis may be overlooked. Only recently has the hydration of olefins by dilute acid been shown to involve general, rather than specific, acid catalysis.

7.2 Specific and general base catalysis

Analogously, specific base catalysis of a reaction occurs in aqueous media when OH^- is the sole catalyst,

$$\text{Rate} = k[OH^-]\ [\text{substrate}], \hspace{3cm} \textbf{2.10}$$

whereas general base catalysis would involve the participation of all the bases present in solution, including the solvent. Water has been chosen as the solvent in these examples of acid and base catalysis, but the distinctions still exist in other media. In general, specific acid catalysis will be shown when only the lyonium ion (the solvated proton, e.g. H_3O^+, $MeOH_2^+$, $EtOH_2^+$) is involved, and specific base catalysis involves only the conjugate base of the solvent or lyate ion (OH^-, OMe^-, OEt^-, etc.).

8 Acidity

8.1 Aqueous systems

When nitric acid is dissolved in water, we show it to be an acid by measuring the H_3O^+ concentration in solution, or by using an indicator. In either case, acidity is shown by the tendency of nitric acid to give a proton to a base (water, or the indicator). The extent to which this reaction occurs depends upon the relative basicities of the acid anion (NO_3^- in this case) and of the solvent. Similarly, the dissociation constant of a base in aqueous solution may be defined in terms of how well the base and OH^- compete for protons. There are two levels past which we cannot accurately measure the acidity or basicity of a substance dissolved in water.

Water, like a number of solvents, undergoes autolysis. Even in the absence of added compounds, pure water contains equilibrium amounts of H_3O^+ and OH^-

$$2H_2O \rightleftharpoons H_3O^+ + OH^-$$

at concentrations of about 10^{-7}M. An acid with a dissociation constant of 10^{-16} would therefore produce an undetectable amount of hydrogen ion even at 1M concentration.

At the other end of the scale, a strong acid (K much greater than unity) may be in equilibrium with its ions, e.g.

$$HNO_3 \rightleftharpoons H^+ + NO_3^-,$$

but if the acid is considerably ionized, the difference between the amount of acid initially added and the amount of hydrogen ion formed is too small to measure. At this situation, all acids appear to be equally strong, and this phenomenon is known as the *levelling effect* of the solvent.

Similarly, all bases beyond a certain strength will appear identical in strength because they will be hydrolysed to form OH^- (e.g. H^-, CH_3^- and NH_2^-).

2.8.2 *Liquid ammonia*

An analogous system to the $H_3O^+/H_2O/OH^-$ system exists in liquid ammonia. All acids show their properties by forming the ammonium ion, and neutralization reactions can be carried out between NH_4^+ and NH_2^- in exactly the same manner as the more familiar analogue in the water system. Because of the greater basicity of the solvent, however, even species such as CH_3CO_2H appear to be strong acids in liquid ammonia. Weak acids in the water system become strong in liquid ammonia, and compounds such as MeOH, $PhNH_2$, and some of the more acidic hydrocarbons appear as weak acids now, although their properties are undetectable in the more acidic water medium.

2.8.3 *Acidic solvents*

Correspondingly, most compounds behave as bases in solvents such as H_2SO_4, liquid HF, or CH_3CO_2H. These media, particularly the first two, tend to protonate most dissolved substances. Nitric acid, for instance, behaves as a base in concentrated sulphuric acid,

$$NO_2OH + 2H_2SO_4 = NO_2^+ + H_3O^+ + 2HSO_4^-,$$

and in acetic acid HNO_3 is a weak acid, with HCl and HBr rather stronger and only $HClO_4$ shows considerable dissociation to give protons to the solvent. In such media we can study the dissociation of acids which were too strong in water to be distinguishable.

2.9 **Acidity functions**

Different solutions of the same acid can show vastly differing apparent acidities, depending upon the solvent. The measurement of acidity of concentrated solutions of acids now presents some problems, for the solution will have its own specific solvent properties. For example, both pure sulphuric acid and $10^{-2}M$ aqueous sulphuric acid have the same hydrogen-ion concentration. Measurement of $[H^+]$ alone cannot reflect the great difference in acidity between the two liquids, and we need a parameter which will measure the proton-donating powers of solutions of acids.

Hammett acidity function H_0

In dilute aqueous media ($\sim 10^{-3}$M) the acidity of a strong acid is entirely due to the concentration of solvated protons ($H_{aq.}^+$; H_3O^+, $H_9O_4^+$, etc.). In concentrated solutions, the effective concentration of a species is not the same as its stoichiometric concentration. The effective concentration (a) is called the *activity*, and is linked to the formal concentration by a factor called the activity coefficient: $a_X = [X] \gamma_X$. In dilute solution, γ tends to unity. If an indicator X is dissolved in such a solution, so that it becomes partially protonated

$$H^+ + X \rightleftharpoons HX^+,$$

an equilibrium constant for the protonation of X may be written

$$k_X = \frac{a_{H^+} a_X}{a_{HX^+}}. \tag{2.11}$$

$$= a_{H^+} \frac{[X]}{[HX^+]} \frac{\gamma_X}{\gamma_{HX^+}}. \tag{2.12}$$

If [X] and [HX$^+$] can be measured (usually spectrophotometrically) then K_X may be derived if the solutions are sufficiently dilute that the activities and the concentrations of the species are effectively the same. Further measurements may now be made, using more concentrated solutions of acid; since [HX$^+$] and [X] are measurable and, since K_X is known, the term $a_H + \gamma_X/\gamma_{HX^+}$ can be obtained. Since the base X will not be useful above acidities where [HX$^+$]/ [X] $\simeq$ 10 (there are difficulties in measuring the concentrations accurately), a less basic indicator Y is now used.

If values of the ratio [HY$^+$]/[Y] are now measured in solutions for which $a_{H^+} \gamma_X/\gamma_{HX^+}$ are known, K_Y can be evaluated, provided that the ratio of activity coefficients is the same for all neutral bases, i.e.

$$\frac{\gamma_A}{\gamma_{HA^+}} = \frac{\gamma_B}{\gamma_{HB^+}} = \frac{\gamma_C}{\gamma_{HC^+}} = \frac{\gamma_D}{\gamma_{HD^+}} = \text{etc.}$$

Experimentally this is so (since K_Y is numerically the same regardless of which base we used to define $a_{H^+} \gamma_A/\gamma_{HA^+}$) and so we may use the derived value of K_Y, and experimentally found values of [HY$^+$]/[Y], to measure the acidities of more acidic solutions. When Y is no longer useful, the process is repeated with Z, an even less basic species.

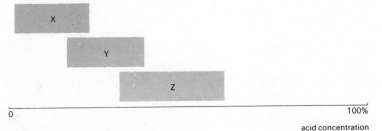

Figure 17 Range over which indicators X, Y and Z are useful

In all cases, the term measured includes a_{H^+}, the true acidity (i.e. the *activity* of the proton-donating species in solution), and an activity coefficient ratio which, in any one solution, is invariant and independent upon the base involved. We cannot separate the two sets of terms, but logically we can define the term h_0,

$$h_0 = a_{H^+} \frac{\gamma_X}{\gamma_{HX^+}},$$ 2.13

which is going to parallel the true acidity of the solution towards any neutral base dissolved in it.

Note that a_{H^+} does *not* refer solely to solvated protons, but is a composite function involving the tendency of all proton-donating species in solution to act as acids. These need not be present in the same concentrations in two solutions of the same h_0; it is not guaranteed that even the same species will be in solution. Thus, one solution may contain H^+Cl^-, HCl, and H_2Cl_2 as well as the various forms of solvated proton; another solution might contain H_2SO_4, HSO_4^-, $H_3SO_4^+$, and perhaps $H_2S_2O_7$ as well as H_3O^+ and $H_9O_4^+$. Both could, however, show the same acidity on the Hammett scale, and the strength of this reasoning lies in the irrelevance of the components of the solution; the only essential is the measure of how well the solution will give a proton to a neutral base dissolved in it, and it is this which h_0 measures.

At low acidities, h_0 tends to equal $[H^+]$; a negative logarithmic form is also used,

$$H_0 = -\log_{10} \frac{a_{H^+} \gamma_X}{\gamma_{HX^+}}.$$ 2.14

There is an obvious analogy with pH, and the two terms tend to equality at high dilutions.

The acidity function therefore parallels $[H^+]$ or pH at low acidities, but carries on through more concentrated solutions of acid in which neither $[H^+]$ nor pH truly reflect the properties of the medium. H_0 has the value of -12 in 100 per cent sulphuric acid, and -11 in anhydrous HF; that is, pure sulphuric acid has one million million times the proton-donating powers of 1M aqueous sulphuric acid.

Other types of acidity function may be similarly defined and measured for bases B^- and B^+; due to the activity coefficient terms, these allied H-functions will not be numerically equal to H_0.

2.9.2 *Gold's function J_0*

The Hammett equation applies to a simple base being protonated by the sequence
$B + H^+ \rightleftharpoons BH^+$,
but another class of neutral bases, the tertiary alcohols, cannot be included in this category, since they ionize in acidic media to give carbonium ions

$$Ar_3COH + H^+ \rightleftharpoons Ar_3C^+ + H_2O,$$ 2.15

as the protonated alcohol $Ar_3COH_2^+$ is unstable. Water, as a weak base, would be protonated to some extent in the acidic solution; since this extent is unknown it is convenient to regard it as unprotonated for the present purposes. The equilibrium constant for the reaction (2.15) is

$$K_{ROH} = \frac{a_{R^+}\, a_{H_2O}}{a_{ROH}\, a_{H^+}} = \frac{[R^+]}{[ROH]}\, \frac{\gamma_{R^+}}{\gamma_{ROH}}\, a_{H_2O}\, \frac{1}{a_{H^+}}. \qquad \textbf{2.16}$$

The Hammett acidity function would correctly represent the acidity of the medium towards the formation of ROH_2^+; i.e.

$$h_0 = \frac{a_{H^+}\, \gamma_{ROH_2^+}}{\gamma_{ROH}}. \qquad \textbf{2.17}$$

Combining equations 2.16 and 2.17 gives

$$K_{ROH} = \frac{[R^+]}{[ROH]} \cdot \frac{\gamma_{R^+}}{\gamma_{ROH_2^+}}\, a_{H_2O}\, \frac{1}{h_0}. \qquad \textbf{2.18}$$

If the ratio of these two activity coefficients in equation 2.18 is unity, then

$$pK_{ROH} + \log_{10}\frac{[R^+]}{[ROH]} = -H_0 - \log_{10} a_{H_2O}.$$

As the activity of water in such solutions can be measured, and as H_0 can be determined, the variations between $\log_{10}([R^+]/[ROH])$ and $H_0 + \log_{10} a_{H_2O}$ for any ROH species which ionizes in this way should be linear. This was found to be true, to within 15 per cent (the slope was 1·16, the discrepancy being attributed to errors in H_0 measurements). A new acidity function was therefore proposed,

$$J_0 = H_0 + \log_{10} a_{H_2O}, \qquad \textbf{2.19}$$

to reflect the acidity of media towards secondary bases (those which lost water during protonation, equation 2.15). Due to the low activity of water in concentrated solutions of acids, J_0 and H_0 have different numerical values. For instance, in 100 per cent H_2SO_4 $J_0 \simeq -19$, while $H_0 \simeq -12$.

2.9.3 Deno's C_0 function

The J_0 function depends upon the assumption that the activity coefficients of R^+ and ROH_2^+ are the same. If this is *not* true, a more complex equation must be used. From equations 2.16 and 2.17 we now obtain the formula

$$pK_{ROH} + \log_{10}\frac{[R^+]}{[ROH]} = H_0 - \log_{10} a_{H_2O} + \log_{10}\frac{\gamma_{R^+}\, \gamma_B}{\gamma_{ROH}\, \gamma_{BH^+}}.$$
$$= C_0.$$

From this, the difference between C_0 and J_0 can be simply expressed as

$$C_0 - J_0 = \log_{10}\frac{\gamma_{R^+}}{\gamma_{ROH_2^+}}. \qquad \textbf{2.20}$$

2.9.4 *Limitations of acidity functions*

The principle of acidity functions requires independence of the value of the acidity function with the nature of the base used. Hammett found that H_0 was appreciably constant for the same solution when a number of allied bases were used; however, the same solution does show distinctly different values of the acidity function depending upon the base employed, and different H values have been measured, depending upon whether a series of anilines, tertiary amines, heterocyclic compounds or non-nitrogenous bases (ketones, olefins, etc.) are used. These values usually lie between those of H_0 and J_0, but they indicate an obvious flaw in the concept.

Bunnett and Olsen have recently proposed a linear free-energy relation which deals more or less successfully with this difficulty. Using the combined term $(H_0 + \log_{10}[H^+])$ as a measure of acidity, which is effectively comparing the reaction under study with the extent of protonation of a hypothetical base of $pK_a = 0$, a number of advantages were evident. Firstly, all the alternative acidity functions varied linearly with this proposed composite acidity function. Secondly, the pK_a of a base could be derived from the intercept of such a plot, taking $\log_{10}([BH^+]/[B])$ as the abscissa. Finally, the acid-dependence of two reactions could be depicted by a linear relation, whether they involved simple protonation,

$$B + H^+ \rightleftharpoons BH^+,$$

in the equilibrium or in the slow stage of the reaction, or whether they involved rapid decomposition of the initial protonated intermediate,

$$ROH + H^+ \rightleftharpoons R^+.$$

The treatment has been satisfactorily applied to 200 reactions, involving both equilibria and kinetic measurements, and fails with only a few. It may well be of great potential.

2.10 Hammett linear free-energy relationship

A number of attempts have been made to rationalize the effects of substituents upon the rate of a reaction or the position of an equilibrium. One of the most successful of these has been due to Hammett. We have seen (section 1.5) that a substituent can affect the dissociation constant of an acid or phenol, due to electronic interactions between the O—H bond responsible for the acidity of the species and the substituent. The dissociation constant of the acid is therefore a measure of the substituent effect in this particular reaction. A linear free-energy relationship links $\log_{10}(k_X/k_H)$ rates of reaction of the substituted and unsubstituted species) or $\log_{10}(K_X/K_H)$ (equilibrium positions of the substituted and unsubstituted species) for two or more processes. Since $\ln k = -RT \Delta F$, the effect of the substituent (measured by k_X/k_H or by K_X/K_H) can be described in changes in the free energy of the process. A linear free-energy relationship simply requires that the effect of a substituent upon two different reactions always occurs to the same extent and in the same direction, although the reaction itself may be more or less sensitive to these influences.

Hammett took the dissociation constants of the benzoic acids, in water at 25 °C, as a standard measure of substituent effects. In postulating a general equation

$$\log_{10} \frac{k_X}{k_H} = \sigma\rho \qquad\qquad 2.21$$

in which σ was the substituent constant (reflecting *only* effects due to the substituent, X, in a defined position) and ρ was the reaction constant (reflecting only the susceptibility of the reaction to substituent effects), he took $\rho = 1\cdot00$ for the dissociation constants of the benzoic acids. This gave a number of substituent constants for substituents in the *ortho*, *meta*, and *para* positions of benzoic acid. If a linear free-energy relationship existed, a plot of $\log_{10}(k_X/k_H)$ against σ for some other reaction (e.g. hydrolysis of substituted ethyl benzoates by alkali in aqueous solution) ought to be linear, and of slope ρ. A linear relationship is found for *meta* and *para* substituents, but not for *ortho* substituents or for aliphatic systems (Figure 18).

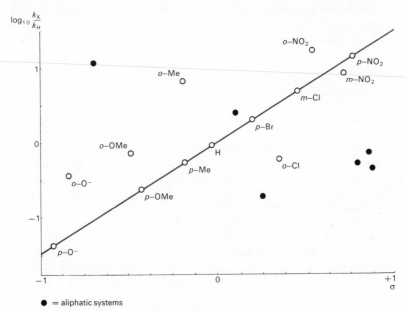

● = aliphatic systems

Figure 18 Hammett free-energy relationship

On reflection, one would not expect *ortho* substituents to show identical effects in different reactions, since the substituent is close to the reaction site where it might exert either a steric effect (which would depend, to some extent, upon the size of the attacking group and upon its orientation) or a direct field effect upon a

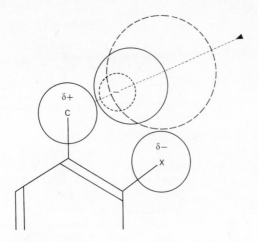

Figure 19 *Ortho* interaction between substituent and reaction site

charged reagent. Since these two effects would be superimposed upon the electronic effects transmitted through the ring, and since their magnitudes would depend upon the nature of the reagent, the observed substituent 'constant' would vary with the reaction under study.

2.10.1 *Significance of the substituent and reaction constants*

Since the dissociation of benzoic acid involves the formation of the negatively charged $ArCO_2^-$ species, a substituent which stabilizes this ion will increase the dissociation constant of the parent acid and will therefore show a positive σ-value; substituents which decrease the dissociation constant similarly must have negative σ. A reaction which is more susceptible to substituent effects, and which also involves the formation of a negatively charged intermediate, will have a value of ρ greater than +1; a process which involves the formation of a *positively* charged intermediate will show a negative value of ρ (e.g. the rates of solvolysis of substituted $PhCMe_2Cl$ derivatives in 90 per cent aqueous acetone shows $\rho = -4\cdot5$; molecular chlorination in nitromethane shows $\rho = -11$).

A low value of ρ (less than unity) is indicative, but not diagnostic, of a homo-lytic reaction mechanism, for here the ease of stabilization of ionic species by substituents (which is what the reaction constant measures) is far less important.

2.10.2 *Limitations of the Hammett equation*

The Hammett equation does not seem to apply to reactions in which relatively high charges are developed. This was first found in a number of solvolysis reactions in which the partial formation of a carbonium ionic intermediate occurred in the rate-determining stage:

$$RCl \xrightarrow{\text{slow}} R^+ + Cl^-. \qquad 2.22$$

Although a good straight line was found with all *meta* substituents and with some (less efficient) *para* substituents, the *para* substituents such as —OMe and —SMe did not fall upon the line. New σ-values could be obtained by moving the experimental results on to the best straight line:

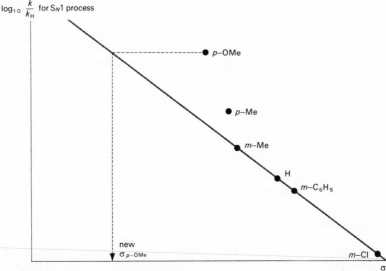

Figure 20 Derivation of the new σ (σ^+) value

but these were not found to apply to other solvolysis reactions. Two treatments of the problem have been suggested.

2.10.3 *Brown σ^+ values*

H. C. Brown studied substituent effects in the solvolysis of some aryldimethyl carbinyl chlorides ($ArCMe_2Cl$) in 90 per cent aqueous acetone at 25 °C, under conditions (equation **2.22**) in which the maximum stabilizing (conjugative) effect of *para* substituents might be realized. A Hammett plot of the results for the *meta* substituents (using σ) gave a good straight line; the experimental results for the other substituents were then fitted to this line to give new values of sigma, σ^+. These values were applicable to reactions in which charge was developed in the transition state, and so could be successfully used not only in solvolysis processes but also in a number of aromatic substitution reactions (see Chapter 6) in which the intermediate possesses some positive charge.

2.10.4 *Yukawa–Tsuno treatment*

On the assumption that the dissociation constants of the benzoic acids reflected a process in which very little conjugative relay occurred between reaction site and substituent, and that Brown's reaction conditions measured the maximum M and

I effects possible from the substituent, Yukawa proposed the equation

$$\log_{10} \frac{k_X}{k_H} = \rho[\sigma + r(\sigma^+ - \sigma)]. \tag{2.23}$$

This simply splits the substituent constant into two factors, one of which (σ) might reflect the inductive effect of the group while the other ($\sigma^+ - \sigma$) could measure the conjugative effect of the substituent; the extent to which conjugation occurs is contained in the proportionality factor r. This refinement gives some idea of the extent to which conjugative stabilization is involved in the transition state of the reaction being studied, but offers very little help in establishing any refinements in a mechanistic study.

2.11   Determination of a reaction mechanism

At this stage, it seems appropriate to consider ways in which one could set about determining the mechanism of a reaction. While not all possibilities can be covered, the following suggestions should be helpful for research purposes.

2.11.1   *Stoichiometry of the reaction and analysis*

We need to know exactly which reaction is under study, and so the first essential is to determine the nature of the reagents, the reaction products, the side-products, and to find out how many molecules of each are consumed and formed in these different processes. It seems obvious that a kinetic study should not begin without an investigation to show whether the supposed reaction is indeed going on, but this requirement is not always met.

To find out exactly in what proportions the reagents interact, and the yields of the products, we must have an effective and quantitative means of analysis.

Inorganic species (e.g. halide ion, bases or acids) may often be analysed by titration or by gravimetric methods; the latter is less common because of the greater inconvenience and longer time required. For instance, the reaction

$$OMe^- + CH_3(CH_2)_3Cl \rightarrow CH_3(CH_2)_3OMe + Cl^-$$

can be conveniently followed either by titration of the base (OMe^-) by standard acid, or by titration of the halide ion produced. At some stage, the changes in concentration of both entities would need to be followed, in order to prove that the reaction followed the stoichiometry of the written equation.

The organic entities, however, cannot usually be identified and assayed so simply, although aldehydes and ketones can often be analysed by conversion to the oximes,

$$RCOR' + H_2NOH.HCl \rightarrow RR'C=NOH + H_2O + HCl,$$

and titration of the strong acid simultaneously formed, while amines and phenols can often be determined by titration with standard acid or base respectively. Phenols and olefins may similarly be determined by titration with standard bromine solutions, or by some similar process involving the ready attack of these species by halogen. In general, these methods require a considerable amount of preliminary work defining the limits of the analytical method and the exact stoichiometry of the process, and are not applicable to all members within the group. For instance, the dinitroanilines and some chlorophenols are too weak to be satisfactorily titrated in aqueous solutions, while acetophenone cannot be analysed through the oxime, and some of the less reactive phenols (e.g. *p*-nitrophenol) require special conditions before they may be analysed by methods involving halogens.

Spectroscopic methods are often useful in identifying the true natures of the reagents and of the products in the reaction solution. This might be necessary if the first product of the reaction were too unstable to survive isolation, or if, in an acidic medium, a reaction proceeded through the protonated form of the original reagent. The various spectroscopic methods are distinguished by the energy of the radiation used; each has its own distinct use.

Ultraviolet and visible spectroscopy can be used to observe electronic transitions in the molecules; the organic reagent and product often differ appreciably in their absorption of ultraviolet and visible light, and we can usually find a position of the spectrum in which all, or almost all, of the light is absorbed by only one species. This point can therefore be used to analyse the concentration of this particular molecule, and hence analyse mixtures containing this species. For example, the cyclization of 2-phenylbenzoic acid to give fluorenone can be plainly followed by the development of the yellow colour of the ketone, as well as the changes in the ultraviolet spectrum of the mixture

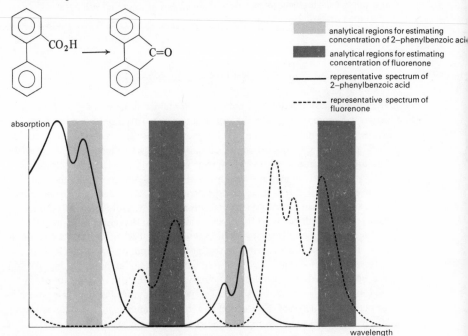

Figure 21 Ultraviolet spectrum of a 2-phenylbenzoic acid-fluorenone mixture

As with other spectroscopic methods, one way of showing the absence of by-products is to compare the u.v. spectrum of a reaction mixture with that of a mixture of reagent and product(s) in the calculated concentrations. Differences in intensity of absorption will be due to the presence of other species, the spectrum of which can be derived by difference.

Infrared spectroscopy can be used analytically as well. The problems here are physical ones, since the extinction coefficients (intensities of absorption of light by the organic molecule) are much less intense than in u.v. spectra, and only a limited number of solvents may be used or the spectrum of the dissolved compound is lost due to the strong blanketing effect of the solvent molecules. However the advantage of i.r. spectroscopy lies in the ease of identification of distinct organic fragments (e.g. C=O, O—H, N—H and benzene substitution patterns). An example of this is the analysis of the concentrations of isomeric methyl biphenyls formed by the attack of toluene by the phenyl radical (Ph·),

$$Ph· + PhMe \rightarrow 2\text{-}, 3\text{- and } 4\text{-}CH_3C_6H_4Ph.$$

Nuclear magnetic resonance (n.m.r.) spectroscopy, and particularly proton n.m.r. or p.m.r. spectroscopy, has also wide application in organic chemistry. Although the method is in principle limited, it can be successfully applied to ^1_1H. As a result, we can see the environment of any proton in an organic molecule, and often identify simple compounds completely. In a typical situation, the product of treating resorcinol (*m*-dihydroxybenzene) with an excess of dimethyl sulphate and base should be *m*-dimethoxybenzene.

The product was kinetically impure, for it contained some species which reacted very much more readily than the main component. This impurity seemed to be an isomer, for it could not be removed by fractional distillation. No impurity could be found by i.r. spectroscopy, but proton n.m.r. showed a main component with four aromatic protons, two of which were equivalent, and six aliphatic protons corresponding to the methoxyl groups in *m*-dimethoxybenzene, together with an isomeric compound with six methyl protons (as in CMe_2), four olefinic protons and a suggestion of a hydroxyl group. This impurity, 2,2-dimethyl-3-hydroxy-3,5-cyclohexadienone, arose from carbon alkylation; the n.m.r. spectrum of *m*-dimethoxybenzene made by another route did not show the presence of any of these absorptions and showed only the 'pure' spectrum of the aromatic ether.

Here the spectrum not only indicated the presence of an impurity, but allowed it to be identified; it is also possible to measure the amount of impurity by integrating the spectrum. In using n.m.r. spectroscopy, we can therefore determine (a) the stoichiometry of the reaction and (b) the amounts and often the natures of the side-products in favourable conditions. Kinetic measurements of the formation of all products would then be intrinsically possible.

Gas chromatography (also called gas–liquid phase chromatography, gas–liquid partition chromatography, and vapour-phase chromatography) is another invaluable means of analysing complex mixtures. In principle it involves the partitioning of a species between a flowing gas phase and a solid phase impregnated with an oil of very low vapour pressure (e.g. silicone oil, dinonyl phthalate) under the temperatures of the separation. The partition ratio depends upon the temperature, the nature of the stationary phase, and uniquely upon the nature of the species on the column. Each component has its own partition ratio, and hence each is retarded to a specific extent as it is partitioned between the gas flow and the inert phase on passing through the column. Complex mixtures can therefore be separated at least partially (gas chromatography effects better separation than some very efficient fractionating columns, for instance), and usually all components may be resolved by using two different columns, where the likelihood of two different species being affected in exactly the same way is small.

Other variations do not involve radically different theoretical considerations.

As each component emerges from the column, it must be detected. This can be done by measuring differences in the thermal conductivity of the gas (taking the entrant gas stream as a reference), or by detecting the presence of organic molecules through a flame-ionization detector or an argon detector. In all cases the response of the detector to the concentration of eluted material is linear only over a range of concentrations; it is always worthwhile calibrating the instrument used in order to be able to measure concentrations, and especially concentration *ratios*, accurately.

In practice, 10^{-6} litre can be separated on an analytical column; the components may be identified by measuring their retention volumes (which depend upon the rate of gas flow and the time between injection and elution from the column) and by comparison with anticipated products. Ideally, this comparison should be made at two temperatures and with two columns in order to show the identities of the component and the added species. For instance, if a natural oil was thought to contain isoeugenol, and one of the peaks from gas chromatography was increased in size on adding this phenol to a sample of the oil, there would be good evidence for supposing the presence of isoeugenol in the natural oil. Only when we could not separate the added phenol from the component in the oil by changes in the chromatography conditions could we be sure of the identity of the component.

Another advantage of gas chromatography is that if the components can be separated, they can be collected. This process, preparative gas chromatography, allows any component to be isolated and characterized in principle. In the previous example, if isoeugenol were not available so that its behaviour on gas chromatography could be studied, the natural oil could be separated by preparative gas chromatography, and the appropriate fraction diverted through cold traps ($-78\ ^\circ$C, with solid CO_2) and the condensed solid then identified by conventional analytical methods.

2.11.2 Means of following the reaction course

We have dealt with some means of analysing a reaction mixture in full. In following the kinetics of a process, we may wish to concentrate only upon one process, and also to use an analytical method which will measure the course of that process to within very small limits. To use some of the spectroscopic devices described would not give sufficient accuracy (unless, of course, there was no other means of analysis) for normal kinetic measurements, and some highly accurate methods are described briefly under this heading.

Ultraviolet and visible spectroscopy is usually capable of the highest required accuracy (i.e. giving rate coefficients which are reproducible to within ±3 per cent and consistent to within ±1 per cent). Most conventional titration techniques (acid-base, or halide ion, for example) are also quite satisfactory. If we wish to refine our measurements still further, it is worth remembering that 1 per cent error in the rate coefficient is brought about by an error or difference of 0·1 °C for many reactions, and that this represents the limit of accuracy of most commercially available thermostats. In principle, both titration methods and spectroscopic techniques (u.v. and visible, not i.r. or n.m.r.) are capable of greater than 1 per cent accuracy.

Other changes can be used to follow the course of the reaction. Measuring a volume of gas evolved (e.g. N_2 from a decomposing diazonium-ion solution) could be a very accurate means of analysis, limited only by the solubility of the gas in the reaction solution and by the limits of the thermostat. In the particular example chosen, it also allows us to concentrate upon those reactions in which nitrogen is evolved; coupling reactions would not be detected in this way and could not interfere with the analysis. The changes in electrical conductivity of a solution also give an accurate means of following the course of a reaction. Again, in the formation of phenol from ArN_2^+ the diazonium ion is exchanged (from a conductance point of view) for a hydrogen ion,

$$ArN_2^+ + H_2O \rightarrow ArOH + N_2 + H^+,$$

and the resulting change is readily measured.

2.11.3 Kinetic analysis

Having found a means of following the course of the reaction, we now need to know which molecules are involved in the rate-determining stage. This can be done by trial and error, fitting the experimental results to successive rate equations; thus, if the reaction A + B = C proceeds at the same rate regardless of the concentration of B, but is first order with respect to A (i.e. rate = $k(a - x)$), then only one molecule of A is involved in the slow stage of the process. In practice, three kinetic studies would be made. In these the relative initial concentrations of the two species could be a and b, a and $2b$, and $2a$ and b. If the reaction were first order with respect to A only, then

$$k_1 t = 2 \cdot 303 \log_{10} \frac{a}{a - x},$$ 2.24

whereas if both A and B were involved in the slow stage

$$k_2 t = \frac{2 \cdot 303}{b - a} \log_{10} \frac{a(b - x)}{b(a - x)}.$$

2.25

If, in all three cases, a good straight line is obtained over about 70 per cent of the reaction course when the appropriate logarithmic term is plotted against time, and if the derived rate constants are indeed constant, there is good evidence for the order of the reaction. If the rate constant changes, some other kinetic equation must be tried (linear plots may result even when the wrong kinetic equation is used!).

Alternatively, a half-life method may be used to determine the order of the reaction separately for each reagent. If two kinetic studies are made in which the concentrations of all the reagents excepting one are kept high and constant (e.g. [B] = 1·0 M with [A] = 0·05 M and 0·10 M respectively), and if the time for each reaction to reach x per cent (x may be 50, but need not be) is measured, then the two times t' and t'' are related to the two initial concentrations (a' and a'') by the relation

$$\frac{t'}{t''} = \left(\frac{a''}{a'} \right)^{z - 1} \quad \text{(where } z = \text{order with respect to A)},$$

from which z may be found. Similarly, two kinetic studies using an excess of A and varying the concentration of B will give the order with respect to B, and so the total order of the process may be found.

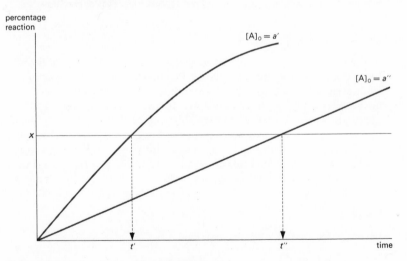

Figure 22 Reaction order with respect to A

A heterolytic mechanism will be more sensitive to substituent effects than will a homolytic process; they are usually also more sensitive to changes in the solvent, but less sensitive to trace impurities. The next stage is to determine whether the reaction is appreciably altered by substituents, whether it is affected by changes in the polarity of the solvent (e.g. from 90 per cent aqueous acetone to 50 per cent aqueous acetone), and whether it is influenced by the acidity or basicity of the solvent (e.g. if the introductory experiments were carried out using $2N\ H_2SO_4$, how dependent is the reaction upon these acidity conditions?). Correspondingly, what are the effects of (i) free-radical initiators, such as benzoyl peroxide or u.v. light, (ii) inhibitors such as iodine, oxygen, hydroquinone, or diphenylamine, (iii) diverting radicals by reaction with scavengers such as 'galvinoxyl' (2.3) or

(2.3)'galvinoxyl'   (2.4) DPPH

diphenylpicrylhydrazyl (DPPH, 2.4) and (iv) adding ions? Some of these experiments will give a clue to the reaction under study; however, it is possible that the addition of a free-radical initiator, or a polar catalyst such as a Lewis acid, may alter the reaction mechanism, and so any results should be viewed with caution. Oxygen, for instance, may initiate and not repress some radical reactions.

2.11.5 *The mechanism of the reaction*

At this stage, the variety of reaction mechanisms becomes so great that there is no general course of procedure. It is now that the definition of the reaction mechanism can start. Essentially, this means not only proving a proposed mechanism (which must be consistent with *all* the experimental results) but disproving, wherever possible, reasonable alternatives. Probably the only test of a mechanism is its consistency with experimental fact; this also means checking that the predictions of the mechanism are true. It must be emphasized that a reaction mechanism is a conclusion based on experimental results; no experimental facts should be neglected in constructing this mechanism, tempting though it might be!

Problems

2.1 Devise means of following the reactions:

(a) $PhCH=CHCH=CH_2 + Br_2 \xrightarrow[25\,°C]{HOAc}$

(b) $PhCH=CHCH=CH_2 + \underset{CH-CO}{\overset{CH-CO}{\underset{\|}{\big|}}} O \xrightarrow[100\,°C]{xylene}$

(c) $PhCH=CHCH=CH_2$ $\xrightarrow{Pd/C}$ 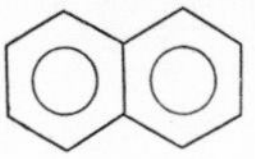 $+ H_2$

(d) Ph_3CCl $\underset{MeNO_2}{\rightleftharpoons}$ $Ph_3C^+ + Cl^-$

(e)

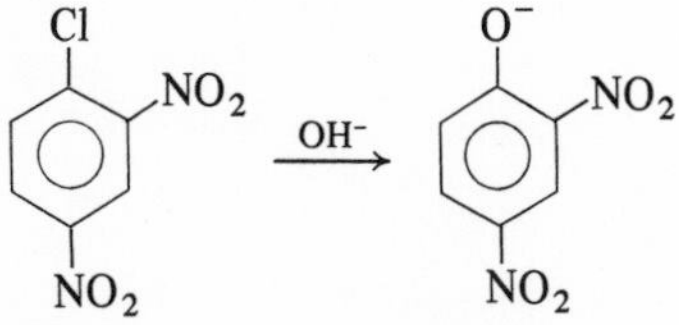

$\xrightarrow{OH^-}$

(f) $ArN_2^+ Cl^- \longrightarrow ArCl + N_2$

2   The rate of the reaction

$$HOCl + ArH \, (+H^+) \rightarrow ArCl + H_2O$$

is increased by increasing concentrations of acid. How could one show whether the true reagent was Cl^+ or $ClOH_2^+$? What would be the effect of using hydrochloric acid as the catalyst? How could one ensure that this side-reaction is not occurring?

3   The rate of solvolysis of diphenylmethyl chloride (Ph_2CHCl) in ethanol is unchanged in the presence of sodium ethoxide, is increased in the presence of sodium nitrate and sodium azide (by about 15 per cent) and is decreased in the presence of lithium chloride. The rate of solvolysis of methyl iodide is increased by all of these salts. Construct a mechanism for each process explaining the observations.

4   The function H_- applies to the protonation of a base A^-. Define this function analogously to H_0, and suggest why it need not parallel the rates of some base-catalysed reactions.

5   In the ionization of some acids in water at 25 °C, Hammett's $\rho = 1\cdot00$ for benzoic acids, but $0\cdot489$ for phenylacetic acids and $0\cdot212$ for β-phenylpropionic acids. Explain the order and direction of these values.

6   Aromatic substitution was thought to involve addition to the aromatic ring, followed by elimination. Evidence for such a proposal included (a) the reaction sequence

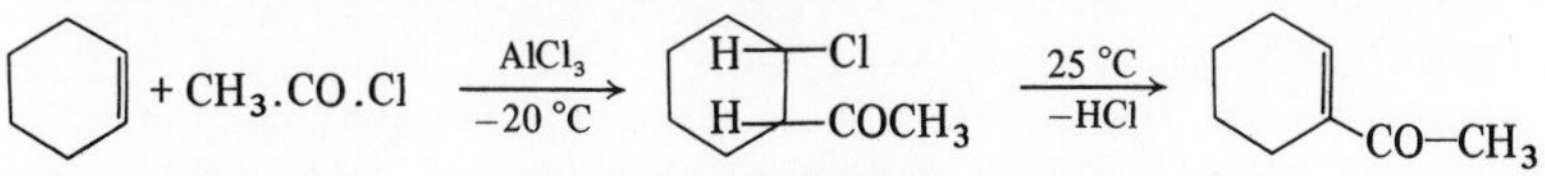

and (b) the isolation of 9,10-dibromo-9,10-dihydrophenanthrene and 9-bromophenanthrene from the reaction of bromine with phenanthrene.

Suggest experiments to disprove this (incorrect) hypothesis.

Chapter 3
Carbonium Ions

3.1 Structure of carbonium ions

We have seen (Chapter 1) that a carbon atom bonded to four other atoms (e.g. CCl_4) has an outer shell of eight electrons. Considering one such covalent bond (e.g. C—X) we have also seen that this bond may break homolytically or heterolytically (section 2.1.1). If the two electrons of the C—X bond remain with the *carbon* atom, a heterolysis occurs to give a *carbanion* (e.g. R_3C^- from R_3CH). If a homolytic process occurs, one electron of the bond stays with each atom and a carbon *radical* results with seven outer-shell electrons. Finally, if both bonding electrons remain with the group X, the heterolysis of the C—X bond gives a *carbonium ion* with six outer-shell electrons (e.g. R_3C^+ from R_3CBr).

Carbonium ions are more commonly found as intermediates than carbanions, although these anions are not rare or insignificant. In both cases, the charged carbon atom is stabilized by spreading the charge over the bulk of the ion; unless considerable stabilization can be achieved, these species are so unstable that they have only transient identity.

Evidence for the existence of some carbonium ions has been available for fifty years; only in especial cases could these intermediates actually be isolated, although recently the identity and nature of a number of *simple* carbonium ions have been studied by n.m.r. techniques. Since carbonium ionic intermediates appear so frequently in our reaction mechanisms, it is appropriate to review the evidence for these structures.

3.2 Evidence for carbonium ionic species

3.2.1 *Physical evidence*

A carbonium ion, since it contains an electron-deficient carbon atom, will be extremely vulnerable to nucleophilic attack. Even the more stable carbonium ions will react readily in solvents such as ethanol or water, either with dissolved anions or with the solvent itself; unstabilized carbonium ions, such as CH_3^+, can usually be detected only in the gas phase where they are unable to attack any other species readily through a sheer concentration effect.

The triphenylmethyl carbonium ion, Ph_3C^+, has ten contributing canonical structures which imply some degree of stabilization (section 1.3). Evidence that this ion could exist in appreciable concentrations came from observations such as the following.

(a) The freezing-point depression of sulphuric acid by triphenylmethanol is nearly four times that produced by an equimolecular amount of some inert solute, such as sulphuryl chloride. It seems that each molecule of the alcohol provides four species on solution, an observation which can be rationalized by the equation

$$Ph_3COH \xrightarrow{H_2SO_4} Ph_3C^+ + H_3O^+ + 2HSO_4^{2-}.$$

(b) Triphenylmethyl chloride and bromide, dissolved in liquid sulphur dioxide, give solutions which conduct electricity as well as solutions of KI of similar concentrations, and whose u.v. spectrum is almost exactly the same as that of triphenylmethanol in sulphuric acid, and quite different to that of the parent halides.

(c) The triphenylmethyl halides, together with some other secondary and tertiary halides, solvolyse by the S_N1 mechanism (Chapter 4) in which the slow stage of the process is thought to involve the ionization of the C—Cl or C—Br bond to form such a carbonium ion.

While this gave good reason to accept the existence of Ph_3C^+, and the isolation of salts such as $Ph_3C^+BF_4^-$ and the corresponding perchlorate strengthened the evidence considerably, it is not always possible to isolate the carbonium ion in this way. Often the existence of these compounds must be inferred from chemical behaviour or, as in recent reports, from spectroscopic measurements. We have seen how n.m.r. spectroscopy allows the identification of organic species as well as their detection, and so we shall not give details of the characterization of simple aliphatic carbonium ions.

.2.2 *Chemical evidence*

In cases where carbonium ionic species are not isolated there are still ways of showing their existence. In some cases this may be from a logistic argument, an example of which is the S_N1 mechanism (Chapter 4). Here only the alkyl halide is involved in the slow stage of the reaction, which is unaffected by the nature or the concentration of the nucleophile. The suggested process is that the alkyl halide *slowly* forms some species which reacts rapidly with any nucleophile available,

$$R—X \xrightarrow{slow} \text{very reactive entity,}$$

and since the alkyl halide itself must react with the nucleophile more slowly than it forms this entity (otherwise the kinetic conditions would not be fulfilled), it follows that the reactive entity must be more electrophilic, i.e. more electron-poor than R—X. The carbonium ion R^+ is an obvious choice which is confirmed by other requirements of the mechanism (e.g. good solvating solvent) and other observations (e.g. stereochemical effects).

Similarly, the addition of bromine to ethylene occurs rapidly in water to give both ethylene bromohydrin and ethylene dibromide (1,2-dibromoethane). In the presence of bromide ion, chloride ion, or nitrate ion the corresponding derivatives of 2-bromoethanol (ethylene bromohydrin) were formed, although this alcohol could not be esterified under the reaction conditions.

$$CH_2=CH_2 \xrightarrow{Br_2} \text{intermediate}$$

with the intermediate reacting via:
- $\xrightarrow{H_2O}$ $BrCH_2CH_2OH$
- $\xrightarrow{Br^-}$ $BrCH_2CH_2Br$
- $\xrightarrow{Cl^-}$ $BrCH_2CH_2Cl$
- $\xrightarrow{NO_3^-}$ $BrCH_2CH_2ONO_2$

Again, the existence of an intermediate with strong electrophilic properties is deduced, and one satisfactory suggestion is the structure $(CH_2CH_2Br)^+$ in which charge is situated mainly or entirely upon the one carbon atom.

3.3 Reactions of carbonium ions and allied species

3.3.1 Reactions with nucleophiles

We have already seen that the electron-poor carbonium ionic centre is readily attacked by nucleophiles. It follows that a carbonium ion will be most likely to survive in an environment of low nucleophilicity and good solvating powers. The stability of Ph_3C^+ in sulphuric acid, nitromethane or liquid sulphur dioxide has been mentioned already; the simpler, and less intrinsically stable, ions are formed in bulk in 'super-acid' media such as $HF–SbF_5–FSO_3H$ mixtures. It should also be pointed out that branching at the α-carbon atom increases the selectivity of the carbonium ion towards nucleophiles because it increases the stability of the ion. This order of stability can be summed up in the series

$$CH_3^+ < CH_2R^+ < CHR_2^+ < CR_3^+.$$

(Note that stabilization by resonance effects may cause inversion of this sequence.)
    It is unlikely that all nucleophiles would react at the same rate with a particular carbonium ion, no matter how intrinsically reactive it might appear to be.

3.3.2 Elimination

An elimination process generally involves the production of a multiple bond by the loss of some small group from a molecule; examples are

$$RCHICH_2I \rightarrow R.CH:CH_2 + I_2,$$
and $R.CH_2.CH_2.OH \rightarrow R.CH:CH_2 + H_2O.$

The process is discussed in more detail in Chapter 5.
    Carbonium ions, which are often intermediates in such reactions, can lose a proton to form olefins; this formally involves an elimination of HX from the carbonium-ion precursor (RX, say). One of the side-products from an alkyl halide undergoing an S_N1 decomposition process is the corresponding olefin; for instance, the solvolysis of t-butyl chloride in ethanol gives isobutylene as well as the corresponding ether (equation **3.1**).

$$(CH_3)_3CCl \xrightarrow{S_N1} (CH_3)_2\overset{+}{C}CH_3 \xrightarrow{EtOH} (CH_3)_3COEt$$

$$\searrow_{-H^+} \quad (CH_3)_2C=CH_2$$

3.1

Such reactions are discussed in more detail in the appropriate chapters; they occur when the carbonium ion is generated in the absence of potent nucleophiles and in the presence of relatively strong bases (e.g. ethanol, water).

,4 Rearrangements

In addition to the simple reactions we have described, the carbonium ion may also rearrange. There are many possibilities within this heading, and the next sections will consider the most important rearrangement processes.

,4.1 *Wagner–Meerwein rearrangements*

This process, originally observed in a number of terpene derivatives, involves the migration of an aryl or alkyl group, or of a hydrogen atom, to the initial carbonium ionic centre so that a new ion is formed. Examples of this would be the isolation of *t*-amyl derivatives from the solvolysis of neopentyl (2,2-dimethyl-1-propyl) halides (equation **3.2**)

$$Me_3CCH_2X \xrightarrow{-X^-} (Me_3CCH_2)^+ \longrightarrow (Me_2\overset{+}{C}CH_2Me) \xrightarrow{Y^-} Me_2CYCH_2Me \quad 3.2$$

and the corresponding reaction of the 2,2,2-triphenylethyl halides, when a phenyl group migrates (equation **3.3**)

$$\underset{\underset{Ph}{|}}{\overset{\overset{Ph}{|}}{Ph-C-CH_2X}} \longrightarrow (Ph_3CCH_2)^+ \longrightarrow (Ph_2\overset{+}{C}CH_2Ph) \xrightarrow{Y^-} Ph_2CYCH_2Ph \quad 3.3$$

In this case, there is evidence that the phenyl group migrates at about the same time as the X group departs; in other words the two ions shown in equation **3.3**, and especially the primary carbonium ion, may not be intermediates but may have merely transitory existences.

3.4.2 *Neighbouring-group participation*

Another aspect of the rearrangement reactions of carbonium ions, and one which is discussed more fully in the next chapter, involves participation by an adjacent group near the carbonium ionic centre. The solvolysis of the 2,2,2-triphenylethyl halides is an example, for here it seems that the carbonium ion is formed with the direct help of the adjacent and migrating phenyl group (see equation **3.3**). Another instance is the solvolysis of the β-chloroamine shown in equation **3.4**.

$$\text{EtCHClCH}_2\text{NEt}_2 \rightarrow \text{EtCH}\underset{\overset{|}{\text{NEt}_2^+}}{\overline{}}\text{CH}_2 \rightarrow \text{EtCH(NEt}_2)\text{CH}_2\text{OH}$$

$$\text{and EtCHOHCH}_2\text{NEt}_2$$

3.4

Both carbon atoms of the ethyleniminium ion intermediate are attacked by nucleophiles, giving a mixture of the rearranged and unrearranged products.

3.4.3 *Pinacol rearrangements*

The Pinacol rearrangement (equation 3.5) and the corresponding reaction with a

$$\text{R}_2\text{C(OH)C(OH)R}_2 \underset{}{\overset{\text{H}^+}{\rightleftharpoons}} \text{R}_2\text{C(OH}_2^+)\text{C(OH)R}_2 \rightarrow \text{R}_2\overset{+}{\text{C}}\text{C(OH)R}_2$$
$$\downarrow \qquad\qquad 3.5$$
$$\text{R}_3\text{CCOR} + \text{H}^+ \leftarrow \text{R}_3\text{C}\overset{+}{\text{C}}(\text{OH})\text{R}$$

β-aminoalcohol, when treatment with nitrous acid gives the intermediate ion $\text{R}_2\overset{+}{\text{C}}\text{C(OH)R}_2$ through the unstable aliphatic diazonium ion (equation 3.6)

$$\text{R}_2\text{C(NH}_2)\text{C(OH)R}_2 \xrightarrow{\text{HONO}} \text{R}_2\text{C(N}_2^+)\text{C(OH)R}_2 \rightarrow \text{R}_2\overset{+}{\text{C}}\text{C(OH)R}_2 + \text{N}_2$$
$$\downarrow \qquad\qquad 3.6$$
$$\text{R}_3\text{CCOR} + \text{H}^+ \leftarrow \text{R}_3\text{C}\overset{+}{\text{C}}(\text{OH})\text{R}$$

both involve the rearrangement of a carbonium ionic intermediate with the migration of the group R. In cases when different aryl or alkyl groups are theoretically capable of migrating (as in the pinacol ArRC(OH)C(OH)RAr, for instance) which group migrates depends upon two general effects. One is stereochemical: the migration of any group in an ion requires that a particular configuration between the migrating group, the leaving group and the carbon chain between them must be established. In compound (3.1) below, R and X must be in a *trans*-diaxial configuration before migration can occur (see also Chapter 4).

$$\begin{array}{c} \text{R} \\ | \\ \text{C}_\beta \!\!-\!\!-\!\!-\!\! \text{C}_\alpha \\ | \\ \text{X} \end{array}$$

(3.1)

If rotation around the C_α–C_β bond is hindered, so that one particular conformation is favoured, this may determine which group in fact migrates. The second, and perhaps more predictable, factor is the availability of electrons from the migrating species to form the bond, because the migration to C_α involves both the bond electrons being supplied by the incoming group. The *p*-anisyl (*p*-MeOC$_6$H$_4$—) group therefore migrates more readily than the *p*-tolyl entity, and an approximate scale of migratory ability can be drawn up considering these factors. This, and many other rearrangement processes, are discussed in Chapter 10.

Allylic rearrangements

The continuity of the π-electron cloud of any conjugated system gives rise to another source of carbonium-ion rearrangement. When the carbonium ionic centre is capable of interacting with a π-system, a number of possible structures of various stabilities may be produced. As a result, nucleophilic attack of the carbonium ion occurs at carbon atoms other than the one which was originally electron-deficient. This situation occurs in the addition of one molecule of bromine to 1,3-butadiene (equation 3.7)

$$CH_2=CHCH=CH_2 + Br_2 \rightarrow BrCH_2CH\overset{+}{\cdots}CH\cdots CH_2 + Br^-$$

$$\downarrow Br^-$$

$$BrCH_2CHBrCH=CH_2$$
$$+$$
$$Br.CH_2.CH:CH.CH_2Br \qquad\qquad 3.7$$

and also in the reactions of some allyl halides (equation **3.8**)

$$RCH=CHCH_2Cl \xrightarrow{-Cl^-} RCH\overset{+}{\cdots}CH\cdots CH_2 + Cl^-$$

$$\downarrow Y^- \text{ (nucleophile)}$$

$$RCHYCH=CH_2 \text{ and } RCH=CHCH_2Y \qquad\qquad 3.8$$

3.5 Carbanions

We began the chapter with a reminder that carbanions exist and are quite stable, under suitable conditions. Most of the reactions we have considered to occur with R_3C^+ have equally as feasible analogies with R_3C^-. The loss of H^-, corresponding to elimination, is a relatively rare occurrence, mainly through the energetics of the process. Rearrangement reactions involving electron-deficient neighbouring groups, or delocalization through conjugated systems, are equally possible but are not as readily observed, mainly because of the relative difficulty in generating carbanions and in finding conditions where they may be attacked by electrophiles other than H^+. It should be recognized, however, that the distinction is an experimental, not a thermodynamic, one.

Problems

3.1   A solution of 1,3-dimethyl-1,3-cyclopentadiene in $FSO_3H–SbF_5$ shows an n.m.r. spectrum with three peaks in the relative intensities of $1:4:6$.

(a) To what is this spectrum due?

(b) What n.m.r. spectrum would be shown by a solution of this hydrocarbon in strongly basic media (e.g. $LiNEt_2 - Et_2NH$)?

(See Deno *et al.*, 1962 and 1963.)

3.2   The hydrolysis of β-(9-anthracenyl)ethyl *p*-toluenesulphonate gives mainly (85 per cent) an isomeric alcohol, whose n.m.r. spectrum in $FSO_3H - SbF_5$ shows four *equivalent* methylene protons. Explain the isomerization and identify the product. (See Eberson, Winstein *et al.*, 1965.)

3.3   The unusually rapid hydrolysis of (chloromethyl)cyclopropane gives the corresponding alcohol (48 per cent), cyclobutanol (47 per cent) and 3-buten-1-ol (about 5 per cent). The same products are formed in the same relative quantities in the hydrolysis of $CH_2=CHCH_2CH_2OTos$ or in the treatment of (cyclopropylmethyl)-amine with nitrous acid. What are the implications of these results? (See Roberts *et al.*, 1951, 1959 and 1964.)

Chapter 4
Substitution at a Saturated Carbon Atom

4.1   Mechanisms of substitution at a saturated carbon atom

4.1.1   *Kinetic differences*

In a study of the rates of solvolysis of alkyl halides in aqueous ethanol or aqueous acetone, two types of kinetic behaviour become apparent. Some halides react with a wide variety of nucleophiles by a kinetic process whose slow stage does not involve the concentration or nature of the nucleophile at all (equation **4.1**),

Rate = k[RX].   **4.1**

Another group of halides are found which react with many nucleophiles by a bimolecular process in which, expectedly, the nature and concentration of the reagent plays an integral part (equation **4.2**),

Rate = k[RX] [nucleophile].   **4.2**

A third, and much smaller, group of halides show behaviour intermediate between these two; the extent to which the nature of the nucleophile is involved seems to vary with the nature of the nucleophile, and appears in the kinetic equation in a fractional order term such as $[Y^-]^{0.4}$. The value of this fraction is also dependent upon the solvent, and the simplest way of regarding these intermediate halides is in terms of a combination of both the general kinetic processes described above.

There are other distinctions between the two main types of reaction. In the first, the presence of other nucleophiles (apart from the solvent) has no effect upon the reaction rate, but the reaction products incorporate the new nucleophile, sometimes almost exclusively. In the second, other nucleophiles accelerate the process remarkably, and the product of the reaction almost entirely incorporates the new nucleophile. These differences led to the postulation of two distinct reaction mechanisms.

4.1.2   *Unimolecular and bimolecular mechanisms*

The kinetic form

Rate = k[RX]

implied that, apart from solvent molecules, only one species was involved in the rate-determining stage. Although solvent attack of RX will give the same kinetic form (rate $= k_2$ [solvent] [RX] $= k_1$ [RX]) the *rate* of such a *bimolecular* process will alter if other nucleophiles are added. The formation of the products appeared to occur in some fast process, and so a unimolecular substitution mechanism (S_N1) was suggested in which the slow stage involved heterolysis of the C—X bond, followed by a rapid reaction between the resulting carbonium ionic intermediate and any nucleophiles present in solution,

$$RX \xrightarrow{\text{slow}} R^+ + X^-,$$

$$R^+ + Y^- \xrightarrow{\text{fast}} RY.$$

The second-order kinetic form,

Rate $= k[RX]$ [nucleophile],

was presumably a one-stage process in which the reagent and the alkyl halide formed the products immediately through one intermediate species. This was a bimolecular substitution process (S_N2); two molecules are involved in the slow stage of the process even though the reaction may show first-order kinetics (as when the reagent is the solvent). The two mechanisms have different requirements and may be readily distinguished. Other mechanisms have been proposed.

4.1.3   *Alternative mechanisms*

The two proposed mechanisms focus attention solely upon the formal reagents, and ignore the role of the solvent (although in fact this plays a vital role; see section 4.1.4). Alternative proposed mechanisms have usually directed attention to the solvent and attempted to explain all reaction processes in terms of a single mechanism. Swain's termolecular mechanism has already been mentioned (section 2.4.3); in this, one or two molecules of the solvent are thought to be involved specifically with each reacting molecule.

In this theory, an alkyl halide reacting by the S_N1 mechanism should show the kinetic behaviour

Rate $= k[RX] = k_3$ [solvent]^2 [RX]

and so k and k_3 should be directly proportional. If a solvent mixture consisting of a very good and a very poor solvent pair (e.g. dioxan–water) were used, an equation such as

$\log k = \log k_3 + 2 \log [H_2O]$

could be set up; $\log k$ ought to vary with the solvent composition. In fact, such variations *are* observed, but they have not always depended upon *twice* the log [H$_2$O], as is required by the trimolecular mechanism. On the other hand, we would expect a simple relation between the composition of the solvent and its solvating properties, particularly at one end of the mixture range, and so a simple relation between this property and k might be expected. There is no compelling evidence that this relation should not be a free-energy one (involving log k) and there is no reason to prefer the proportionality factor of two against any other figure.

The difference between these two explanations is that the Swain theory (which, with variations, has been advanced by other workers periodically) involves the solvent molecules kinetically; the Ingold theory states that the solvent molecules are thermodynamically involved: they help the reaction to proceed but do not specifically participate.

4.2 Requirements of $S_N 1$ and $S_N 2$ mechanisms

The essential difference between the $S_N 1$ mechanism and the $S_N 2$ process is a matter of timing. For the unimolecular mechanism to occur, the rate of heterolysis of the C—X bond must be greater than the rate of attack of the nucleophile upon the α-carbon atom of the alkyl halide. Factors which will assist the ionization of the halide will facilitate the $S_N 1$ process; those which promote nucleophilic attack upon $C^{\delta +}$—$X^{\delta -}$ will make an $S_N 2$ process more likely. The polarity of the C—X bond will obviously be increased (and so heterolysis is more likely) if X is a strongly electron-withdrawing group (such as halogen), and if R the organic fragment forms a relatively stable carbonium ion and has substituents attached to the α-carbon atom which stabilize and assist the formation of positive charge. Ionization of the alkyl halide will also be greatly assisted by a polar solvent; experimentally it is found that the unimolecular mechanism is far more susceptible than the bimolecular mechanism to changes in the solvent. Lastly, because there is the competition between heterolysis of the C—X bond ($S_N 1$) and direct attack by the nucleophile ($S_N 2$), the unimolecular process is favoured by the absence of highly reactive nucleophiles.

A typical $S_N 1$ reaction, therefore, is the ethanolysis of diphenylmethyl chloride. Chlorine is strongly electronegative, and the $Ph_2 CH$ fragment forms a relatively stable carbonium ion. The solvent is a highly polar one, and the only nucleophiles present in solution are ethanol and, at later stages of the reaction, chloride ion. In contrast, an $S_N 2$ mechanism is found in the reaction of ethyl iodide with methoxide ion in methanol; although the solvent is highly polar, and the halogen atom is electron-withdrawing, Et^+ is not a very stable ion, and OMe^- is a very powerful nucleophile. Not surprisingly, benzyl halides show intermediate behaviour in aqueous acetone and in alcohol. The benzyl carbonium ion has some stability, and while $S_N 2$ processes are found in the reactions with piperidine or with hydroxide ions, mixed order reactions are observed with milder nucleophiles (e.g. solvolysis). In this situation it is uncertain whether the benzyl halide has time to ionize before attack of the nucleophile, and so both $S_N 1$ and $S_N 2$ processes occur simultaneously.

4.2.1 *Steric requirements of the two mechanisms*

An $S_N 2$ displacement results in inversion of the configuration of the α-carbon atom; the name Walden inversion has been given to all such displacements. An example is the formation of l-acetates from d-chlorides, as in equation 4.3.

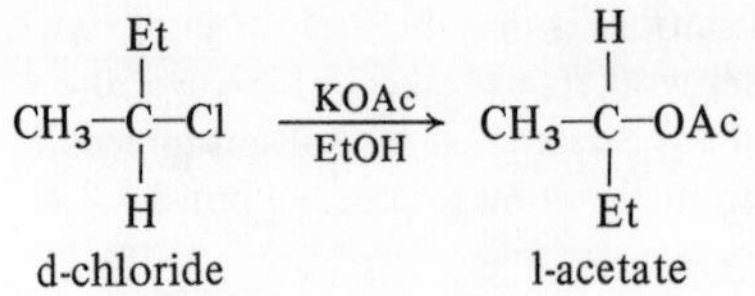

$$\underset{\text{d-chloride}}{\overset{\text{Et}}{\underset{\text{H}}{CH_3-C-Cl}}} \xrightarrow[\text{EtOH}]{\text{KOAc}} \underset{\text{l-acetate}}{\overset{\text{H}}{\underset{\text{Et}}{CH_3-C-OAc}}}$$

The rates of inversion and the rates of nucleophilic attack are the same, implying that the transition state in the S_N2 process must involve distortion of the tetrahedral angles between the α-carbon atom and the groups attached. As S_N2 displacements will only occur when the reagent is able to approach the α-carbon atom in the line of the C–X bond and from the 'back' of the molecule (hence the low reactivity of Me_3CCH_2Cl, where the bulky t-butyl group precludes attack by the nucleophile upon C_α) the course of the displacement must be

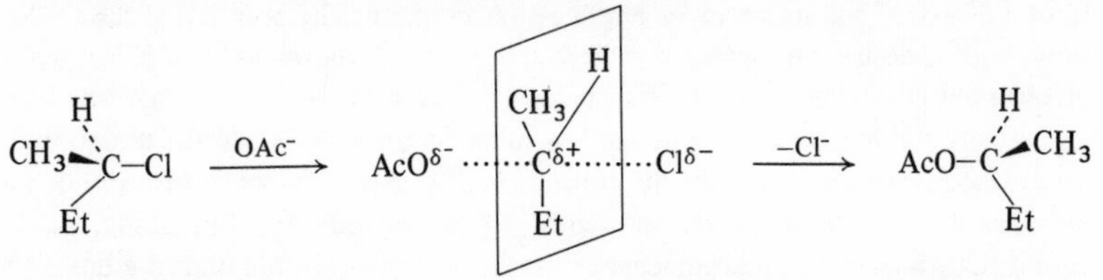

in which the α-carbon atom and the three groups attached become coplanar, while the entering and leaving groups move in directions perpendicular to this plane. The orientation of attacking group and C–X bond has been explained in terms of electrostatic effects; however, such an explanation does not account for the observed inversion processes found in S_N2 reactions of ions such as $RNMe_3^+$ or $RSMe_2^+$.

In contrast with the optical inversion which accompanies S_N2 displacements, the S_N1 mechanism is accompanied by racemization when either d- or l-halides give products with little or no optical activity. Since the rate-determining stage involves the formation of a carbonium ion, a species in which the α-carbon atom has sp^2 character and so becomes coplanar with the three groups attached to it, attack by the nucleophile may take place from either side of this planar ion to form the d- and l-products with equal ease, always provided that the leaving group is sufficiently far away from the carbonium ion.

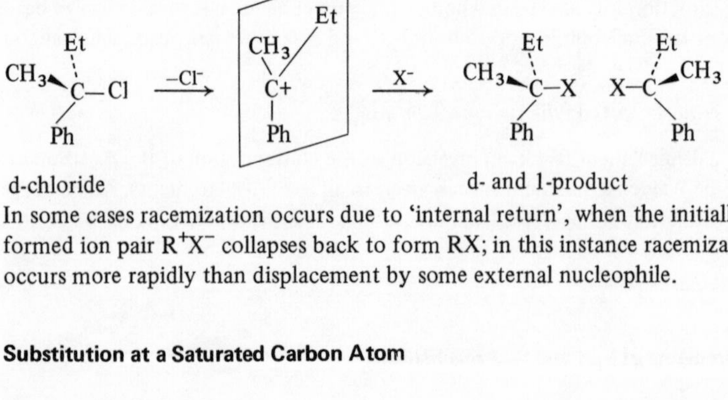

In some cases racemization occurs due to 'internal return', when the initially formed ion pair R^+X^- collapses back to form RX; in this instance racemization occurs more rapidly than displacement by some external nucleophile.

Although the S_N1 mechanism does not require any orientation between reagent and alkyl halide, it does require the formation of a planar carbonium ion. If the three bonds between C_α and the substituents cannot become nearly coplanar, the S_N1 mechanism is inhibited. An extreme example of such behaviour is found in the 'bridge-head' halides.

4.2.2 Non-reactive halides

In a species such as 1-chlorobicyclo[2,2,2]octane (4.1), the S_N2 process cannot occur, for the nucleophile must approach in the line of the C—Cl bond and from the opposite side – which is blocked by the bicyclo-octane structure. (cf. chlorocyclohexane, where approach is hindered but not impossible). The S_N1 mechanism requires the formation of a nearly planar carbonium ion in which the three methylene groups attached to C_α become nearly coplanar with it. Such distortion of the molecule would be strongly resisted, although it represents the only energetically feasible mode of reaction. In contrast to 'normal' alkyl halides, the halogenobicyclo-octanes react only very slowly with ethanolic silver nitrate. The more rigidly held 1-halogenotriptycenes (4.2) are inert to silver ion, sulphide

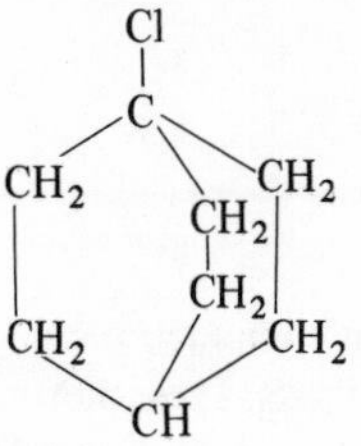

(4.1)

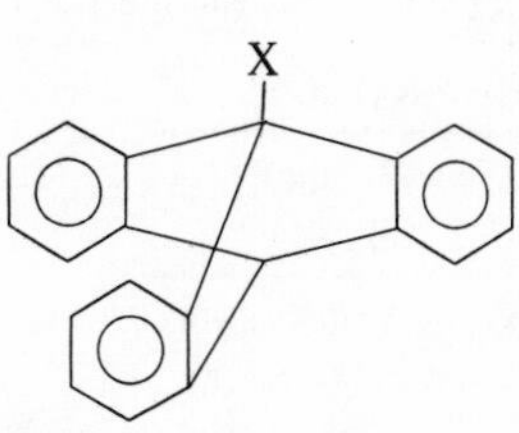

(4.2)

ion (a powerful nucleophile in S_N2 displacements) or hydroxide ion in ethanol at 80 °C, and reaction occurs only slowly at 140 °C. In contrast, the triphenylmethyl halides solvolyse rapidly, by the S_N1 mechanism, in ethanol at 0 °C.

4.2.3 Steric acceleration in S_N1 processes

Since the formation of the carbonium-ion intermediate involves expulsion of a group and the relief of steric strain within the molecule, halides with very bulky substituents attached to C_α may show steric acceleration effects. For example, the two crowded halides

$$
\begin{array}{ccc}
\quad\text{Me} & & \quad i\text{-Pr} \\
\quad| & & \quad| \\
t\text{-Bu}\!-\!\text{C}\!-\!\text{Cl} & \text{and} & i\text{-Pr}\!-\!\text{C}\!-\!\text{Cl} \\
\quad| & & \quad| \\
\quad t\text{-Bu} & & \quad t\text{-Bu}
\end{array}
$$

react some thousands of times as rapidly as t-BuCl. In these cases back-strain (B-strain) is considerable, and the relief of this strain in forming the carbonium ion may well provide some driving force for the reaction; indeed, it has been suggested that some of the reactivity of Ph_3CCl is due to such steric acceleration.

4.2.4 *Steric retardation of S_N2 processes*

In forming the intermediate in an S_N2 mechanism, the nucleophile must approach C_α sufficiently to form a partial bond. This may be extensively hindered by bulky substituents in either the substrate or the reagent. For instance, the stability of $R_3N \to BR_3$ complexes between substituted boranes and 2-substituted pyridines decreases in the order

$$BH_3 > BF_3 > BMe_3$$

for the boranes, and

$$2\text{-Me} > 2\text{-Et} > 2\text{-}i\text{-Pr} > 2\text{-}t\text{-Bu}$$

for the substituted pyridines. This effect is not due to electronic interactions, for the 3- and 4-alkylpyridines do not show such dependence. In the S_N2 reactions of the 2-alkylpyridines with alkyl iodides in nitrobenzene at 25 °C, the relative orders of reactivity are evidently due to steric effects (Table 1) and in fact the inertness

Table 1 Relative Reactivities in the Reaction $R'_3N + RI \to R'_3NR^+I^-$

X-Substituent in pyridine	2-H	2-Me	2-Et	2-i-Pr	2-t-Bu	3- and 4-alkyl
MeI	4290	2025	955	306	1·00	11900–8900
EtI	228	53·3	24·4	6·9	–	540–500
i-PrI	11·7	0·64	–	–	–	25·1–19·5

of 2,6-dimethylpyridine to nucleophilic attack is the basis of most methods for its purification from traces of other isomers. Increasing the size of the substituents in the pyridine system causes a greater susceptibility to the size (steric demands) of the reagent; there is evidence (Brown, 1956) that even transference of a proton to the nitrogen atom of 2,6-di-t-butylpyridine is hindered.

A further indication of steric hindrance in S_N2 processes is shown by the relative reactivities of triethylamine (NEt_3) and quinuclidine (4.3) (1-aza-bicyclo-[2,2,2] octane)

(4.3) quinuclidine

towards isopropyl iodide; the rigidly held amine reacts at seven hundred times the rate of NEt_3, where presumably at least one alkyl group may shield the 'face' of the nitrogen atom and hinder attack.

3   Neighbouring-group participation

The low reactivity of neopentyl halides $(Me_3C.CH_2X)$ has been ascribed to the bulky t-butyl group, which prevents approach of a nucleophile to C_α and so precludes the operation of the S_N2 process. Reaction presumably occurs through some sort of S_N1 mechanism, which will not be favoured because of the relative instability of a primary carbonium ion (sections 3.2.2 and 4.2.1). It is therefore surprising to find that Ph_3CCH_2Cl reacts at several thousand times the rate of the neopentyl halides, although the Ph_3C group is undoubtedly as bulky, if not more so, than the t-butyl group. Both sets of halides give rearranged products,

$$R_3CCH_2X \xrightarrow{\ Y^-\ } R_2C=CHR + R_2C(Y)CH_2R,$$

but it is thought that the phenyl groups may actually assist in the heterolysis of the C—X bond by migrating during the rate-determining stage. This is an example of neighbouring-group participation, a process which was first detected during the solvolysis of optically active α-bromopropionic acid derivatives.

3.1   *Evidence of neighbouring-group participation*

The base-catalysed hydrolysis of d-α-bromopropionate anion normally provides the l-α-hydroxypropionate ion (l-lactate) and, similarly, methyl α-bromopropionate provides methyl lactate with inversion of configuration. This behaviour is typical of an S_N2 displacement (equation 4.4). However, at very low concentrations of OH^-,

$$CH_3-\underset{\underset{\displaystyle H}{|}}{\overset{\overset{\displaystyle Br}{|}}{C}}-CO_2Me \xrightarrow{\ OH^-\ } CH_3-\underset{\underset{\displaystyle OH}{|}}{\overset{\overset{\displaystyle H}{|}}{C}}-CO_2Me \qquad\qquad 4.4$$

d-α-bromopropionate gives d-lactate. The S_N2 mechanism requires *inversion* of configuration, while the S_N1 mechanism allows *racemization*; here is a process involving *retention* of configuration. The process is best explained by postulating initial interaction between the CO_2^- group and the α-carbon atom (equation 4.5) in which the bromine atom is displaced by the near approach of O^-. The resulting

$$CH_3-\underset{\underset{H}{|}}{\overset{\overset{Br}{|}}{C}}-C\overset{\overset{O}{\diagup}}{\underset{O^-}{\diagdown}} \quad \longrightarrow \quad CH_3-\underset{\underset{H}{\vdots}}{\overset{\overset{Br^{\delta-}}{\vdots}}{C^{\delta+}}}-\underset{O^-}{C}=O \qquad \mathbf{4.5}$$

intermediate may be an α-lactone, or may be a carbonium ionic intermediate in which there is considerable interaction between C^+ and O^-:

$$CH_3-\underset{\diagdown O \diagup}{CH-C}=O \quad or \quad CH_3-\underset{\ddots O^-}{\overset{+}{CH}-C}=O$$

There is controversy over the extent to which this carbon–oxygen bond is formed, but it is only necessary that the participating oxygen atom effectively shields C_α from nucleophilic attack from one side. The nucleophile is then only able to attack from the one side of the system, that from which Br^- was expelled.

4.3.2 Rearrangement accompanying neighbouring-group effects

The reactivity of alkyl halides bearing some β-substituents (e.g. $Et_2NCH_2CH_2Cl$, $EtSCH_2CH_2Cl$) is ascribed to neighbouring-group participation (equation 4.6).

$$Et_2N-\underset{\underset{CH_2Cl}{|}}{CH_2} \quad \xrightarrow{-Cl^-} \quad Et_2\overset{+}{N}\underset{\diagdown CH_2 \diagup}{-CH_2} \quad \xrightarrow{Y^-} \quad Et_2NCH_2CH_2Y \qquad \mathbf{4.6}$$

The formation of such three-membered ring systems is shown by the isolation of rearrangement products; thus $EtSCH(Me)CH_2Cl$ provides not only $EtSCH(Me)CH_2OH$ on hydrolysis, but also $EtSCH_2CHOHMe$ (equation 4.7).

$$EtS-\underset{\underset{CH_2Cl}{\diagup}}{CH(Me)} \quad \longrightarrow \quad Et\overset{+}{S}\underset{\diagdown CH_2 \diagup}{-CH(Me)} \quad \xrightarrow{OH^-} \quad \overset{\nearrow EtSCH_2CH(Me)OH}{\underset{\searrow EtSCH(Me)CH_2OH}{}} \qquad \mathbf{4.7}$$

Similar effects are found in which nitrogen, sulphur, oxygen, halogen and even aryl and alkyl groups appear to participate during the slow stage of the reaction to give rearrangement products. Thus in the acetolysis of some derivatives of PhCHMeCHMeOH the L-*threo-** esters give D- and L-*threo-* acetates (equation 4.8),

$$\qquad \mathbf{4.8}$$

whereas the L-*erythro-* ester solvolyses with complete retention of configuration (equation 4.9).

* *Threo-* and *erythro-* refer to the stereochemistry about two adjacent carbon atoms. The *erythro*-isomer is that in which two similar groups (H and Me in the present case) can eclipse each|other on rotation about the $C_\alpha-C_\beta$ bond.

4.9

3.3 Acceleration due to neighbouring-group effects

The relative rates of reaction of a number of β-substituted ethyl halides lie in the order

$EtCH_2CH_2Cl$	HCH_2CH_2Cl	$EtOCH_2CH_2Cl$	$EtSCH_2CH_2Cl$
0·7	1·0	1·6	$\sim10^4$

and similar sorts of accelerations are found in most cases where neighbouring-group participation can occur. In an attempt to measure these effects, Winstein studied the rates of acetolysis of some 2-substituted cyclohexyl esters.

cis   *trans*

Only the *trans*-substituted isomers were able to show neighbouring-group participation (section 4.3.5) and since all electronic effects ought to be comparable between the *cis* and *trans* isomers, any difference in rate between the two ought to be a measure of the neighbouring-group effect. The ratio of the rates was expressed in terms of free-energy changes (cf. section 2.10) called the *driving force G*. This had values ranging from 54 kJ mol^{-1} (SEt) to nearly zero (Cl). Other measurements of driving force, although falling in the same general pattern, were numerically different; this may be because, in the cyclohexyl system, the driving force contains within it the energy difference between the diequatorial and diaxial configurations. The latter, although a higher-energy structure, is the configuration through which neighbouring-group interaction occurs.

3.4 Steric requirements for neighbouring-group participation

We have already suggested that neighbouring-group participation requires a particular orientation of the group and the reaction centre. In fact, these effects only occur when the group, the carbon atoms attached to it and to the leaving group, and the leaving group itself can become coplanar, and then only when the neighbouring group can attack C_α from behind (4.4). It is for this reason that the *trans*-2-substituted cyclohexyl esters only could show neighbouring-group effects. However, *trans*-configuration in itself is not sufficient: there is no difference in rate between the two chlorosulphides (4.5) and (4.6), nor in the corresponding

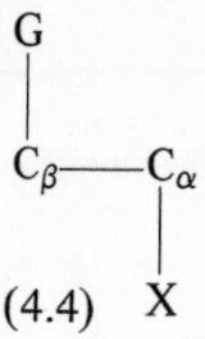

G
|
C_β—C_α
(4.4)   X

H—SAr
Cl
H

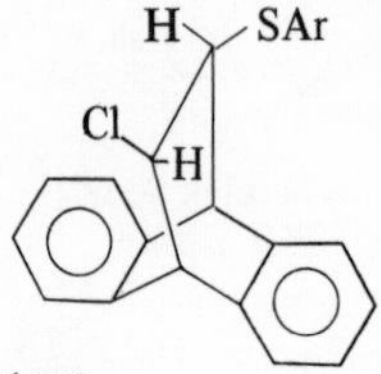

(4.5)

H—SAr
H—Cl

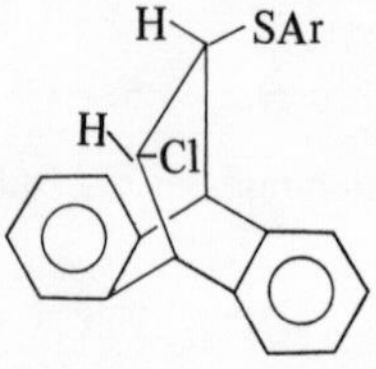

(4.6)

norbornane system (4.7) and (4.8). In these cases the ring structures are too rigid

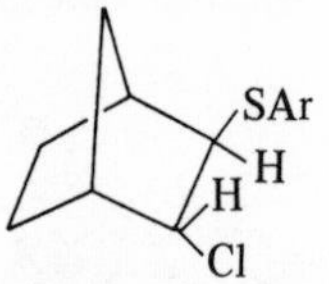

SAr
H
Cl

(4.7)

H
SAr
H
Cl

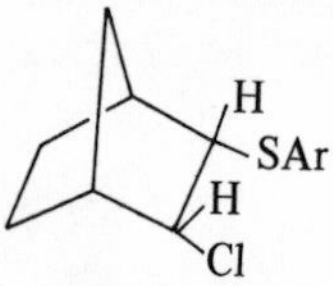

(4.8)

to allow coplanarity between the four participating atoms. These stringent orientation effects are similar to those required in the S_N2 mechanism, and neighbouring-group participation has been described as an internal S_N2 process. This description reflects the stereochemical consequences of the effect (for two bimolecular displacements would give effective retention of configuration) but implies the complete formation of a bond between the group and C_α, which has not been shown in all cases.

Problems

4.1 In the acetolysis of the *erythro*- and *threo*-esters PhCHMeCHMeOTos, show that a cyclic phenonium ion will explain (a) the formation of racemic products from optically active *erythro*-isomers, and (b) the retention of optical activity in the solvolysis of optically active *erythro*-isomers, if attack can occur at either C_α or C_β. (c) Since p-MeOC$_6$H$_4$CHMeCHMeOTos solvolyses at 300 times the rate of the unsubstituted ester, do we have evidence of interaction between the aryl group and C_α? (d) If isomeric esters of p-MeOC$_6$H$_4$CHMeCHMeOH are the *only* products of solvolysis of this tosylate, which of the following phenonium ion structures are precluded?

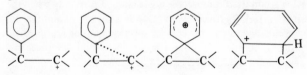

4.2 For the classical S_N1 mechanism

$$RX \underset{k_{-1}}{\overset{k_1}{\rightleftharpoons}} R^+ + X^-$$

$$R^+ + Y^- \xrightarrow{k_2} RY$$

show that a first-order reaction will be observed when $k_{-1}[X]$ is much less than $k_2[Y]$, and so explain the curve in the reaction plot as the reaction progresses.

4.3 The orientation of attack in the S_N2 process was once thought to be due to electrostatic effects. Show that this would explain inversion in the reaction

D-PhCHClMe + PhS$^-$ → L-PhCH(SPh)Me + Cl$^-$

but not in the displacement

D-PhCHMeNMe$_3^+$ + OAc$^-$ → L-PhCHMeOAc + NMe$_3$.

4.4 The hydrolysis of $PhCCl_3$, $PhCHCl_2$, and $PhCH_2Cl$ in 50 per cent aqueous acetone increases in the order $PhCH_2Cl < PhCHCl_2 < PhCCl_3$. Only the reaction of benzyl chloride is accelerated by added OH$^-$, although not to give the kinetic equation rate = $k[PhCH_2Cl][OH^-]$. Explain the results, remembering that α-Cl usually retards S_N2 displacements.

4.5 Explain the order of solvolysis of the following ω-bromoalkyl amines $H_2N(CH_2)_n Br$ in water at 25 °C.

n	2	3	4	5	6
k_{rel}	75	1	62 000	1000	1·2

Chapter 5
Elimination Reactions

5.1   Reactions involved

An elimination reaction involves the splitting out of a small group (H_2O, NMe_3, HX) from an organic molecule, with the formation of a cyclic system or, especially, a multiple bond. Examples are the formation of olefins from alcohols, or from base-catalysed eliminations from esters or alkyl halides, and the formation of cyclo-alkanes; it also includes the dehydration of 1,1-diols (giving aldehydes or ketones) and the formation of nitriles from oximes (equations **5.1–5.4**).

$$RCH(OH)CH_3 \xrightarrow{H^+} RCH=CH_2 + H_2O, \qquad\qquad \textbf{5.1}$$

$$RCH_2CH_2Cl \xrightarrow{OEt^-} RCH=CH_2 + EtOH + Cl^-, \qquad\qquad \textbf{5.2}$$

$$RCH_2CH(OH)_2 \xrightarrow{H^+} RCH_2CHO + H_2O, \qquad\qquad \textbf{5.3}$$

$$RCH=N(OH) \xrightarrow{H^+} RCN + H_2O. \qquad\qquad \textbf{5.4}$$

We shall be particularly concerned with β-eliminations, where the two fragments eliminated come from adjacent carbon atoms.

5.2   Elimination mechanisms

5.2.1   *El mechanism*

Corresponding to the S_N1 and S_N2 mechanisms, there also exist unimolecular and bimolecular elimination processes. In the S_N1 mechanism we have considered possible fates of the carbonium ionic intermediate formed in the slow stage of the reaction; one of these is loss of a proton (equation **5.5**),

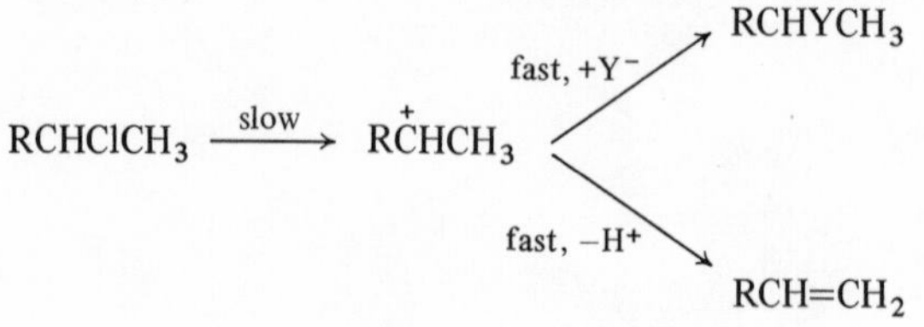

$$RCHClCH_3 \xrightarrow{slow} R\overset{+}{C}HCH_3 \begin{cases} \xrightarrow{fast,\ +Y^-} RCHYCH_3 \\ \\ \xrightarrow{fast,\ -H^+} RCH=CH_2 \end{cases} \qquad \textbf{5.5}$$

and so a unimolecular elimination mechanism is possible. The relative extents to which nucleophilic attack of the carbonium ion competes with proton loss will depend, among other things, upon the relative stabilities of the products. In keeping with this, elimination seems to be more preferred in highly crowded systems where the formation of an olefin involves relief of steric strain. Steric factors are not the only ones, for we must also consider the relative stabilities of the olefins formed. In general, the most highly substituted ethylene is the most stable olefin; hence although n-$PrCMe_2Cl$ is somewhat more crowded than $EtCMe_2Cl$, it forms slightly less olefin because of the relative stabilities of the products (equations **5.6, 5.7**).

$$CH_2CH_2CH_2\overset{+}{C}Me_2 \longrightarrow CH_3CH_2CH{=}CMe_2 \qquad\qquad \textbf{5.6}$$

$$CH_3CH_2\overset{+}{C}Me_2 \longrightarrow MeCH{=}CMe_2 \qquad\qquad \textbf{5.7}$$

The small difference in stability is attributed to hyperconjugative effects. Hyperconjugation (or 'no-bond resonance') is thought to involve stabilization through secondary structures such as

$$CH_3{-}CH{=}CH_2 \longleftrightarrow H^+\,\bar{C}H_2{-}CH{=}CH_2$$

$$\updownarrow$$

$$H^+\,CH_2{=}CH{-}\bar{C}H_2,$$

and so the order of efficiency of stabilization will be CH_3 (three protons participating) $> RCH_2 > R_2CH > R_3C$. It has been suggested that even more remote effects may result from hydrogen attached at β-carbon atoms (e.g. $H{-}C{-}C{-}C{=}C$) but the clear definition of such effects has not been accomplished. While both olefins are trialkyl ethylenes, one of them ($MeCH{=}CMe_2$) is hyperconjugatively stabilized by nine hydrogen atoms and the other has only eight atoms so involved. Hyperconjugation is also thought to explain the relative stabilities of the substituted ethylenes: $R_2C{=}CR_2 > R_2C{=}CHR > RCH{=}CHR > RCH{=}CH_2$. In many cases, however, the elimination products can be more simply anticipated. Acetolysis of $Me_3CCH_2CH(OBs)CH_3$ gives predominantly the disubstituted ethylene.

$$Me_3C{\diagdown}CH_2{-}CH{-}CH_3 \longrightarrow \underset{75\%}{\overset{Me_3C}{\diagdown}{\underset{H}{}}C{=}C\overset{H}{\underset{CH_3}{\diagup}}} + \underset{24\%}{\overset{Me_3C}{\diagdown}CH_2CH{=}CH_2}$$
$$\underset{OBs}{|}$$

Bs = 'brosylate' = p-bromobenzenesulphonate

The *trans* olefin is the major product; the *cis* isomer, in which there is considerable interaction between methyl and t-butyl groups, appears in about 1 per cent yield.

5.2.2 *Bimolecular elimination process*
The analogue to the S_N2 mechanism

Rate = k [base] [alkyl halide]

is a common occurrence in elimination reactions. One possible mechanism involves the formation of equilibrium amounts of the carbanion derived from the alkyl halide, and this can then lose halide ion to give the olefin:

$$\text{Base} + -CH_2CH_2X \rightleftharpoons -\bar{C}HCH_2X + BH^+,$$

$$-\bar{C}HCH_2X \xrightarrow{\text{slow}} -CH=CH_2 + X^-.$$

β-Phenylethyl bromide is a particularly favourable case where the carbanion ought to have reasonable stability; however, no deuterium is incorporated either in the recovered halide or in the styrene formed when the elimination is carried out in D_2O or EtOD. Hence this mechanism does not seem relevant to normal alkyl halides, although it may occur where carbanion formation is thoroughly favoured (e.g. $BrCH_2CH(CO_2Et)_2$).

The most satisfactory mechanism seems to be a concerted process in which the removal of a proton from C_β and the displacement of the leaving group from C_α occur simultaneously (equation 5.8). The stereochemistry of the elimination also seems to involve a *trans*-diaxial disposition of the proton and the X group, and all four atoms again seem to require coplanarity for elimination to take place. In this sense there is an obvious parallel with S_N2 displacements, and it has been suggested that a four-centre transition state (5.1) is formed in each case. When attack at C_α occurs, displacement takes place (S_N2); when proton abstraction prevails, elimination results (E2). As yet there seems no compelling evidence for this simplification, nor has it been shown to be generally applicable.

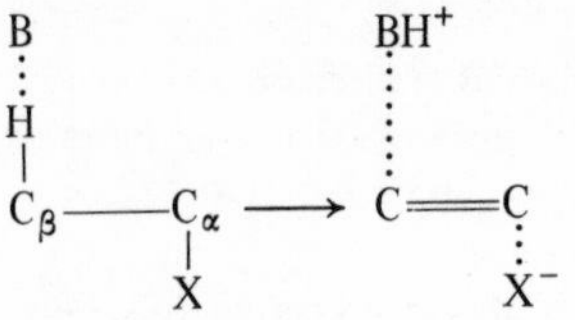

5.8

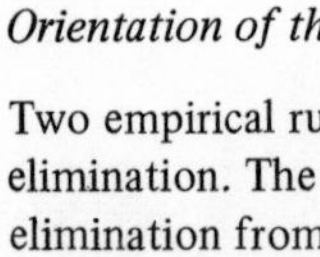

(5.1)

5.2.3 *Orientation of the olefin products from E2 reactions*

Two empirical rules were formulated governing the nature of the olefin formed by elimination. The more substituted ethylene was formed during bimolecular elimination from alkyl halides (Saytzeff rule, e.g. equation 5.9), whereas during elimination from tetra-alkylammonium or tri-alkylsulphonium ions the less-substituted ethylene was the predominant product (Hofmann rule, e.g. equation 5.10). Until recently it was felt that a Hofmann-type reaction product resulted when the acidity of the proton removed was of prime importance, whereas the Saytzeff rule prevailed when the stabilities of the olefin products (and their relevant transition states) were the determining factors. In fact the extent of 'Saytzeff' to 'Hofmann' elimination can be varied either within a series of alkyl

$$RCH_2CHClCH_3 \xrightarrow{\text{Saytzeff}} RCH=CHCH_3 \qquad \textbf{5.9}$$

$$\underset{\underset{+}{NMe_3}}{RCH_2CHCH_3} \xrightarrow{\text{Hofmann}} RCH_2CH=CH_2 \qquad \textbf{5.10}$$

halides, or within a series of bases bringing about the elimination of HX from one alkyl halide, by steric effects. For example, elimination from Me_2CHCMe_2Br gives 80 per cent $Me_2C=CMe_2$ (using OEt^-) down to 9 per cent (using Et_3CO^-), a complete transition from Saytzeff- to Hofmann-type elimination products. As the groups eliminated are much bulkier in compounds showing Hofmann-type elimination, the two types of behaviour can be rationalized in terms of the crowding in the various transition states (**5.2, 5.3**).

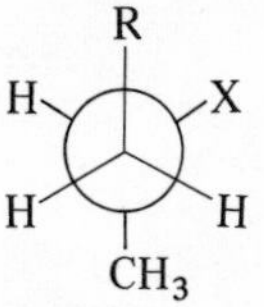

$$RCH_2CHXCH_3$$

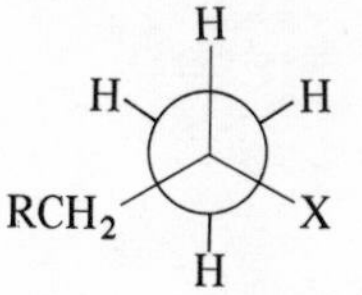

(5.2) Saytzeff   (5.3) Hofmann

Steric interference can arise if either the alkyl group or the leaving group are large, and in order to achieve the *trans*-diaxial conformation a less 'acidic' hydrogen atom may be preferred, when the hindering alkyl group is removed from the sphere of the reaction site.

Problems

5.1   β-Benzene hexachloride (5.4) reacts with alkali at one ten-thousandth of the rate

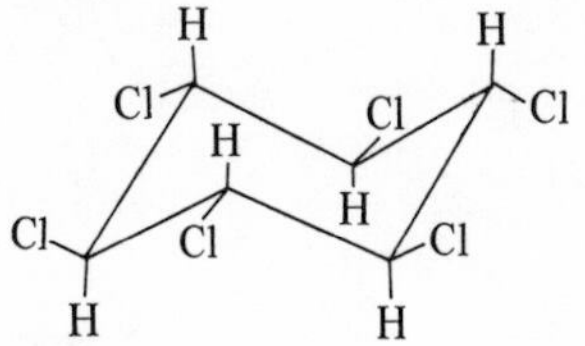

(5.4)

of its conformational isomers. Suggest a reason for this. Also explain how β-benzene hexachloride takes up deuterium when the elimination is carried out in a deuterated solvent. (See Cristol, 1947, 1949, 1951 and 1953.)

5.2   The rates of base-catalysed dehydrohalogenation of some benzaldehyde *N*-chloroimines ($ArCH=NCl \rightarrow ArCN$) in 92.5 per cent EtOH at 0 °C are increased by electron-withdrawing groups in the phenyl system (Hammett $\rho = +2\cdot2$). Substituents with considerable mesomeric effects (e.g. *p*-NO_2) are somewhat more effective than predicted from the Hammett equation. Are these results consistent with an intermediate of the structure $Ar\overline{C}=NCl$? (See Hauser, LeMaistre and Rainsford, 1935.)

5.3   The base-catalysed hydrolysis of DDT (2,2-*bis*(*p*-chlorophenyl)-1,1,1-trichloro-
      ethane) gives the corresponding diarylacetic acid. The relative reactivities of the
      mono- di- and tri-chloroethanes is in the order Ar_2CHCH_2Cl (1·0) $<$ $Ar_2CHCHCl_2$
      (6·0) $<$ Ar_2CHCCl_3 (150). Explain this order, and devise a mechanism which also
      explains the isolation of $Ar_2C=CCl_2$ from the trichloroethane hydrolysis.

5.4   The product of bromine addition to *trans* cinnamic acid loses HBr and CO_2 in
      basic media to give β-bromostyrene. The anion of the dibromo-acid eliminates by a
      first-order process. In aqueous media, *trans* β-bromostyrene constitutes the major
      isomer (78 per cent); in acetone, the *cis* isomer is formed exclusively. In solvents of
      intermediate polarity, the proportion of *cis* to *trans* products varies accordingly.
      Suggest an explanation. (See Grovenstein and Lee 1953 and Cristol and Norris, 1953.)

Chapter 6
Aromatic Electrophilic Substitution

Aromatic electrophilic substitution is one of the most well-studied fields of physical organic chemistry, for it includes nitration, halogenation, Friedel–Crafts acylation and alkylation, sulphonation and a host of other preparative processes. Nitration, in particular, has been the basis of much of our knowledge of substituent effects, and was one of the model reactions used to define and measure mechanistic concepts.

6.1   **Nitration**

The reaction

$$ArH + HNO_3 \rightarrow ArNO_2 + H_2O$$

may be carried out under a number of reaction conditions, the most common of which is in solution in strong acid. While the reaction seems to involve electrophilic attack upon the aromatic system (cf. conditions needed to mono-nitrate PhOH, PhH, $PhNO_2$, and m-dinitrobenzene), the identification of the electrophile is not easy. Early kinetic studies of the nitration of nitrobenzene in sulphuric acid showed the simple form

$$Rate = k[PhNO_2][HNO_3],$$

where the concentrations of each reactant were quoted without consideration of any subsequent process (such as protonation) which might have altered these values from the stoichiometric quantities.

The anomalous behaviour of solutions of nitric acid in sulphuric acid (van't Hoff factor = 4) suggested the formation

$$HNO_3 + 2H_2SO_4 = NO_2^+ + H_3O^+ + 2HSO_4^- \qquad \textbf{6.1}$$

of the nitronium ion (NO_2^+, equation **6.1**); confirmatory evidence was found from measurements of infrared spectra of such solutions, electrolysis and the isolation and X-ray crystallographic studies of salts such as nitronium perchlorate, borofluoride, etc.

6.1.1 *The identity of the electrophile*

Although there is adequate evidence for the presence of NO_2^+ in solutions of nitric acid in concentrated sulphuric acid, other electrophiles, notably $H_2NO_3^+$, exist in such solutions. These species may be distinguished by studying the dependence of the rate of nitration upon the acidity of the reaction medium. The nitric acidium ion, $H_2NO_3^+$, is generated by simple protonation (equation 6.2)

$$HNO_3 \xrightleftharpoons{H^+} H_2NO_3^+ \hspace{3cm} 6.2$$

whereas the nitronium ion results from the loss of water from this first intermediate (equation 6.3).

$$HNO_3 \xrightleftharpoons{H^+} NO_2^+ + H_2O. \hspace{3cm} 6.3$$

The equilibrium concentrations of these two species will vary with the acidity of the medium, but the formation of $H_2NO_3^+$ will parallel the H_0 function $(B + H^+ \rightleftharpoons BH^+$; see section 2.9) and the extent of formation of NO_2^+ will parallel J_0. Since the rate of nitration of nitrobenzene in 90-100 per cent H_2SO_4 was found to parallel the extent of formation of the tris-*p*-nitrophenylmethyl carbonium ion from the parent carbinol, dependency upon J_0, and hence the intermediacy of the nitronium ion, was proven. Since the reaction involved one molecule of nitric acid (or some derivative) in the rate-determining stage, and since the true electrophile was formed by protonation with the loss of water, the true reagent must be the nitronium ion.

6.1.2 *Reactions in nitromethane*

In the presence of a large excess of nitric acid (necessary to minimize interference from the water formed in the reaction, which participates in a number of competing equilibria), the nitration of benzene and of toluene in $MeNO_2$ is a zeroth-order process with respect to the aromatic compound,

Rate $= k$ (or rate $= k'[HNO_3]^n$).

Sulphuric acid increased this rate, and added NO_3^- decreased the rate, without altering the order of the reaction. The rate-determining stage must therefore involve only nitric acid, and not the aromatic species. Less reactive aromatic species showed rates of reaction which were dependent upon the concentration of substrate, until a first-order process

Rate $= k[ArH]$

became evident for the nitration of ethylbenzoate, 1,2,4-trichlorobenzene, and the dichlorobenzenes.

The first stage of the reaction, involving nitric acid alone, is the formation and subsequent dehydration of $H_2NO_3^+$ (equations 6.4, 6.5).

$$2HNO_3 \rightleftharpoons H_2NO_3^+ + NO_3^-, \hspace{3cm} 6.4$$
$$H_2NO_3^+ \rightleftharpoons NO_2^+ + H_2O. \hspace{3cm} 6.5$$

With a highly reactive aromatic substrate (e.g. benzene or the alkylbenzenes), NO_2^+ is removed from the system as soon as it is formed; the rate-determining stage now becomes the rate of formation of the electrophile. This explains the acceleration due to sulphuric acid (stronger acid giving a greater equilibrium concentration of the nitric acidium ion) and the retardation due to NO_3^- (which represses the equilibrium **6.4** and so lowers the equilibrium concentration of $H_2NO_3^+$).

In less-activated aromatic systems, the rate of reaction of NO_2^+ with ArH (**6.6**) does not successfully compete with the back reaction of water with NO_2^+. As k_2 diminishes, the equilibrium between $H_2NO_3^+$, NO_2^+, and H_2O develops until, with

$$\text{ArH} + NO_2^+ \xrightarrow{\ k_2\ } \text{ArNO}_2 + \text{H}^+. \qquad\qquad\qquad \textbf{6.6}$$

ethyl benzoate and other deactivated systems, it is fully developed. The nitration stage now becomes the rate-determining step, and a first-order kinetic process prevails under the reaction conditions.

6.1.3 *Nitration in other solvents*

Acetyl or benzoyl nitrate (nitronium acetate or benzoate) are mild nitrating agents, usually used for polyannular hydrocarbons or non-benzenoid aromatic substitutions. Acetic acid and/or acetic anhydride, or carbon tetrachloride are the usual solvents under these conditions, when it is doubtful whether NO_2^+ itself is the reagent. There is also evidence that such nitration processes are not direct substitution processes, but addition–elimination sequences, for the reaction of indan with acetyl nitrate solutions gives not only 5- and 6-nitroindan, but also acetoxyindans (equation **6.7**).

Section 6.1.4 deals with the mechanism of nitration under these conditions.

Dinitrogen pentoxide (N_2O_5) in carbon tetrachloride has also been studied kinetically. The solid exists as nitronium nitrate ($NO_2^+NO_3^-$) and may be expected to provide NO_2^+; however the kinetics of the process are complicated by autocatalysis from the nitric acid produced during the reaction (**6.8**),

$$N_2O_5 + \text{ArH} \rightarrow \text{ArNO}_2 + \text{HNO}_3, \qquad\qquad\qquad \textbf{6.8}$$

when a whole family of contributions appear in the kinetic equation, of which two at least have been isolated, corresponding to catalysis involving the second and third powers of the nitric acid concentration. Nonetheless, the simple bimolecular process

Rate $= k[\text{ArH}]\,[N_2O_5]$

can be isolated. It is interesting to note that benzoyl nitrate has also been shown to operate not as a nitrating agent *per se,* but through the formation of equilibrium quantities of N_2O_5; here benzoic anhydride represses the observed nitration rates (**6.9**).

$$2PhCOONO_2 \rightleftharpoons (PhCO)_2O + N_2O_5 \qquad\qquad 6.9$$

6.1.4   *Mechanism of nitration by* NO_2^+

The kinetics of the reaction indicate that both NO_2^+ and ArH are involved in the transition state; under conditions in which the formation of this transition state is the slow stage of the sequence, it has been shown that hexadeuterobenzene and benzene react at identical rates, and that the nitration of *o*-tritiotoluene gives *o*-nitrotoluene in which half of the tritium has been displaced. Since the three hydrogen isotopes are replaced at identical rates, the breakage of the C—H bond cannot be the rate-determining stage; the sequence must therefore be

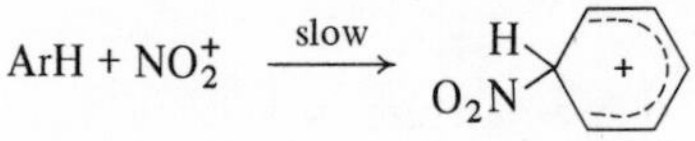

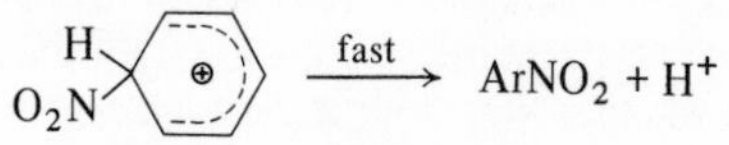

and in keeping with this mechanism, no base-catalysis of nitration has been observed.

This mechanism involves direct *substitution.* The alternative *addition–elimination mechanism* was proposed as a mechanism of aromatic substitution; it was refuted on the basis that, for example, the nitration of ethylene does not give β-nitroethanol (expected if the nitration of benzene to give nitrobenzene involves such similar adducts) but ethyl nitrate (equation **6.10**).

$$CH_2{=}CH_2 + HNO_3 \rightarrow CH_3CH_2ONO_2 \ (not\ CH_2OHCH_2NO_2) \qquad 6.10$$

Similarly, the formation of 9-bromophenanthrene and of 9,10-dibromo-9,10-dihydrophenanthrene were shown to be parallel, not consecutive, processes (**6.11**).

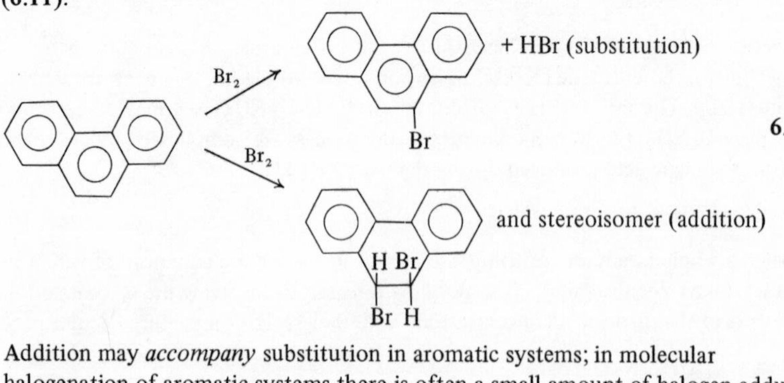

Addition may *accompany* substitution in aromatic systems; in molecular halogenation of aromatic systems there is often a small amount of halogen adduct formed, but these represent a separate mode of reaction and are not precursors to the aromatic halide product.

.1.5 *Nitration through prior nitrosation*

Very reactive species, such as phenol, may be nitrated under conditions where nitronium ion does not exist in any quantity. Under these conditions there is evidence of nitration involving initial nitrosation. Although nitrous acid retards nitration in strongly acid media, presumably because it provides bases (NO_2^- and NO_3^-) to divert NO_2^+ (equations 6.12–14),

$$2HNO_2 \rightleftharpoons N_2O_3 + H_2O \qquad\qquad 6.12$$
$$N_2O_3 \rightleftharpoons NO^+ + NO_2^- \qquad\qquad 6.13$$
$$NO_2^+ + NO_2^- \rightleftharpoons N_2O_4 \rightleftharpoons NO^+ + NO_3^- \qquad\qquad 6.14$$

it accelerates the attack of phenols and phenol ethers under relatively low acid conditions. HNO_2 is a stronger base than HNO_3, and so NO^+ can exist in concentrations of acid in which NO_2^+ cannot. When the aromatic system is sufficiently reactive towards the weaker electrophile (NO^+) and when the medium is not sufficiently strongly acidic to permit kinetically significant concentrations of NO_2^+ to exist, we can imagine nitration proceeding through the initial attack by NO^+, and subsequent oxidation of the nitroso group to $-NO_2$ by nitric acid. In keeping with this, *p*-nitrosophenol has been isolated from such reactions; however, this implies that the oxidation is not as rapid as in the kinetic studies, and it may be that other oxidizing agents rather than HNO_3 intervene.

.2 Sulphonation

The mechanism of the sulphonation reaction has not been so clearly defined. From a study of the sulphonation of $ArNO_2$ and $ArNR_3^+$ species in oleum, the reaction sequence 6.15–17 has been proposed, in which the slow stage of the process is the loss of a proton from the intermediate. In aromatic substitution, since attack by the electrophile and expulsion of the proton occur in two stages, there must be an intermediate in which both groups are bound to the aromatic system, so that the carbon atom under attack acquires sp^3 character. Such an intermediate is called a Wheland intermediate. This mechanism agrees with the positive isotope effect found in the sulphonation of tritiobenzene and tritiobromobenzene, which implies that proton loss from carbon occurs in the rate-determining step.

$$ArH + SO_3 \rightleftharpoons [ArHSO_3] \qquad\qquad 6.15$$

$$[ArHSO_3] + H^+ \rightleftharpoons \left[Ar{<}^H_{SO_3H} \right]^+ \qquad\qquad 6.16$$

$$\left[Ar{<}^H_{SO_3H} \right]^+ \xrightarrow{\text{slow}} ArSO_3H + H^+ \qquad\qquad 6.17$$

In nitromethane, sulphur trioxide reacts by the third-order process,

Rate= $k[\text{ArH}][\text{SO}_3]^2$.

The significance of the second molecule of sulphur trioxide is not certain. It may increase the acidity of the intermediate (cf. the function of the proton in **6.16**); it may be that SO_3 is involved in more than one stage of the sequence; or it may be that the true reagent is S_2O_6.

Any mechanism proposed for sulphonation must take into account the reversibility of the process; this also makes a kinetic study more difficult and has contributed to the uncertainty. Most proposed sequences suggest that proton loss from the Wheland intermediate (6.1) is the rate-determining stage of the process,

$$
\underset{(6.1a)}{\overset{\oplus}{\bigcirc}} \quad \text{H} \quad \overset{\ominus}{SO_3} \qquad \text{or} \qquad \underset{(6.1b)}{\overset{\oplus}{\bigcirc}} \quad \text{H} \quad SO_3H
$$

but the earlier stages of the reaction are not so clearly defined.

6.3   **Halogenation**

6.3.1   *Molecular halogenation*

Halogenation of aromatic systems is an electrophilic process, for aromatic ethers and phenols are more readily attacked than alkylbenzenes, which are in turn much more susceptible to substitution than nitrobenzene or benzoic esters. In polar solvents (MeNO_2, $\text{CH}_3\text{CO}_2\text{H}$) the reaction shows second-order kinetics,

Rate = $k[\text{ArH}][\text{X}_2]$.

The true electrophile appears to be the halogen molecule itself, and not X^+ derived from the pre-equilibrium

$$X_2 \rightleftharpoons X^+ + X^-, \qquad\qquad\qquad\qquad \textbf{6.18}$$

since the reaction rate is not greatly repressed by the addition of X^- ions. (Some repression would be expected, due to the formation of trihalide ions which would decrease $[X_2]$ in solution.)

In the case of chlorination, slight acceleration is observed, when a salt effect (section 2.5) overrides the (small) formation of trihalide ion. Similarly, the suggestion that chlorination in acetic acid involved chlorine acetate (ClOAc) as the true electrophile was disproved (i) by the synthesis and study of ClOAc, which reacts less readily than solutions of molecular chlorine in acetic acid, and (ii) by the positive salt effect of acetate ions, arguing against a heterolysis similar to equation **6.18** and a source of Cl^+.

3.2 Catalysis in halogenation

Lewis acids catalyse molecular halogenation. The effect, first thought to involve an increased polarity of the halogen–halogen bond through intermediates such as $[X^+FeX_4^-]$, seems to indicate assistance in decomposing an initially formed complex between halogen and the aromatic species $(X_2.ArH)$. These catalysed reactions show third-order kinetics,

$$\text{Rate} = k[ArH][X_2][\text{Lewis acid}],$$

but there is no instance where the process becomes zeroth order in aromatic hydrocarbon. This could occur when the rate of heterolysis of the halogen–halogen bond (giving X^+) became the slow stage of the process (cf. nitration in $MeNO_2$) and so seems to argue against the increased polarity of the reagent by the catalyst.

3.3 Hypohalous acids; X^+

The hypohalous acids, particularly HOCl, could act as halogenating agents and ionize in the sense $\overset{\delta+}{X}-\overset{\delta-}{OH}$ (cf. $\overset{\delta+}{X}-\overset{\delta-}{X}$ for the molecular halogens). Since the OH group is a poorer leaving group than X, the hypohalous acids only substitute the more activated aromatic systems such as PhOH and PhO⁻. The reaction shows simple second-order kinetics in the absence of added acid,

$$\text{Rate} = k[HOCl][ArH].$$

The reaction is remarkably catalysed by halide ion, since the more reactive free halogens are then formed

$$2H^+ + Cl^- + OCl^- \rightarrow Cl_2 + H_2O.$$

It is also accelerated by added acid,

$$\text{Rate} = k[ArH][HOCl][H^+]$$

and for the more reactive species, the chlorination becomes zeroth order with respect to aromatic compounds,

$$\text{Rate} = k[HOCl][H^+].$$

This last kinetic form implies that the rate-determining stage involves only the hypohalous acid, and from the stoichiometry of the process it must reflect the formation of either $ClOH_2^+$ or Cl^+. The proton transfer

$$H_3O^+ + ClOH \rightarrow ClOH_2^+ + H_2O$$

is unlikely to be slow enough to limit such a reaction; it seems more feasible that the breakage of the Cl—O bond

$$ClOH_2^+ \rightarrow Cl^+ + H_2O \qquad\qquad 6.19$$

is this slow process. In agreement with this, the reaction proceeds more rapidly in D_2O; although D^+ transfer would be a slower process than H^+ transfer, the lower basicity of D_2O would mean that higher equilibrium concentrations of protonated hypohalous acid would exist in solutions of the same stoichiometric hydrogen-ion concentration (cf. section 2.6). Against this evidence for the intermediacy of Cl^+ is a calculation of the energy of the heterolysis process **6.19** which gives a ridiculously high value for what should be a low-energy process. However, Cl^+ is not being formed in isolation, but in a solvent of unique solvating powers; this may provide sufficient stabilization energy to make the two apparently diverse results consistent.

The kinetic studies are consistent with the mechanism

$$XOH \underset{}{\overset{H^+}{\rightleftharpoons}} XOH_2^+ \rightleftharpoons X^+ + H_2O \xrightarrow{ArH} \left[Ar{<}^{H}_{X} \right]^+ \longrightarrow ArX + H^+,$$

in which either the formation of X^+, or its attack upon the aromatic species, may be the rate-determining stage. No hydrogen isotope effects have been found, and so proton loss cannot be kinetically significant.

6.3.4   *Iodination*

The low reactivity of iodine is reflected by the limited number of studies on its kinetics of aromatic substitution. The iodination of aniline in aqueous iodide shows the kinetic form

$$\text{Rate} = \frac{k[\text{base}][\text{PhNH}_2][I_3^-]}{[I^-]^2}$$

under base catalysis. This form is equivalent to a number of other processes, such as

$$\text{Rate} = k'[\text{PhNH}_2][\text{base}][I^+] \qquad \textbf{6.20a}$$
$$= k''[\text{PhNH}_3^+][\text{base}][\text{HOI}] \qquad \textbf{6.20b}$$
$$= k'''[\text{PhNH}_2][\text{base}:H^+][\text{HOI}]. \qquad \textbf{6.20c}$$

Some of these processes can now be ruled out. A sequence involving $PhNH_3^+$ would be expected to give *meta*-iodoanilines from electrophilic substitution; as *para*-substituted products are obtained, equation **6.20b** is unlikely. The implication of equation **6.20a** is a mechanism similar to that of chlorination and bromination

$$C_6H_5NH_2 + I^+ \rightleftharpoons [C_6H_5NH_2.I]^+ \xrightarrow[\text{slow}]{\text{base}} IC_6H_4NH_2 + BH^+,$$

but in which the decomposition of the Wheland intermediate by base is the slow process. An alternative mechanism, suggested by equation **6.20c**, involves the formation of $[PhNH_2IOH]$ in a fast pre-equilibrium, followed by the slow abstraction of OH^- by BH^+ and the subsequent rapid loss of a proton from the Wheland intermediate.

A similar multiplicity of possible mechanisms is found in the iodination of phenol, where iodine cation I^+ or HOI may be the true electrophile, and PhOH or PhO^- may be the true substrate. In this case, however, the iodination shows a strong isotope effect. 2,4,6-Trideuterophenol is substituted at only one quarter of the rate of phenol. Proton loss from carbon is therefore kinetically significant, and the mechanism seems to be

$$\text{PhOH} + \text{I}^+ \rightleftharpoons \overset{\text{base}}{\underset{\text{slow}}{\rightleftharpoons}} \quad + \text{BH}^+ \qquad \textbf{6.21}$$

By analogy, we would expect the same process to prevail with aniline.

.5 *Mechanism of halogenation*

Halogenation involving molecular *chlorine* or *bromine* is a bimolecular process in polar solvents, and shows no hydrogen isotope effect; the slow stage of the process must therefore be the formation of the Wheland intermediate and not its decomposition (equation **6.22**).

$$\text{X}_2 + \text{ArH} \rightleftharpoons [\text{ArH.X}_2] \overset{\text{slow}}{\rightleftharpoons} \quad \overset{\text{fast}}{\longrightarrow} \text{ArX} + \text{H}^+ \qquad \textbf{6.22}$$

In the case of iodination in basic media, the removal of the proton from the Wheland intermediate is kinetically significant; whether all other stages of the reaction are fast, or whether both the formation and the decomposition of the intermediate are of similar rates, cannot be decided (**6.21**).

Halogenation involving the *hypohalous acids* have, as the rate-determining stage, either the formation of Cl^+, or the attack of the aromatic system by this species, in acid media. In the absence of acid, attack of the aromatic system seems to be the slow stage; there is no evidence of proton loss from the aromatic substrate becoming kinetically significant.

4 Diazonium ion coupling reactions

The formation of dyes by the reaction of diazonium ions with phenols or with amines has been industrially significant for some years. Kinetically, the reaction with phenols is a simple second-order process,

$$\text{Rate} = k[\text{ArO}^-][\text{ArN}_2^+],$$

which again allows more than one mechanistic interpretation. The true substrate may be either the phenol or phenoxide ion, and the true reagent may be ArN_2^+ or $\text{ArN}=\text{NOH}$. The variation of rate of attack of *amines* by diazonium ions with pH indicates that ArN_2^+ is the true electrophile. Consistent with this is the activating effect of electron-withdrawing groups (e.g. *p*-nitrobenzene diazonium ion reacts with anisole, but benzene diazonium ion will not react with phenol, and only with the more reactive phenoxide ion). While in general the reaction shows no hydrogen isotope effect, so that proton-loss from carbon in the Wheland intermediate is not kinetically significant, base-catalysis is found in some cases; these also show a hydrogen isotope effect. The rate of such reactions does not change proportionally with changes in the concentration of base, and a two-stage process has been suggested (6.23).

$$Ar'N_2^+ + ArH \rightleftharpoons \underset{\oplus}{H} \diagdown N=NAr' \xrightarrow[\text{slow}]{\text{base}} ArN=NAr' + BH^+ \qquad 6.23$$

6.5 Friedel–Crafts alkylation and acylation

The alkylation and acylation of aromatic hydrocarbons requires Lewis-acid catalysts; these combine to some extent with the alkyl or acyl halide to form 1:1 complexes. The complexes possess some electrical conductivity, and the halogen atoms in species such as $[RXAlX_3]$ attain some degree of equivalence, for radioactive 'scrambling' occurs between EtBr and radioactive $AlBr_3$. The incipient ionization to provide $R^+AlX_4^-$ cannot be fully developed, however; alkylation reactions in nitrobenzene or trichlorobenzene follow the law

Rate = $k[ArH][RX][AlX_3]$.

If the highly reactive carbonium ion R^+ were formed (and became the true electrophile in the system), the rate of the reaction

$$R^+ + ArH \rightarrow ArR + H^+$$

would then show dependency upon $[AlCl_3]^{\frac{1}{2}}[RX]^{\frac{1}{2}}$, since

$$K = \frac{[R^+][AlX_4^-]}{[RX][AlX_3]} = \frac{[R^+]^2}{[AlX_3][RX]}.$$

In *acylation* reactions, although similar kinetics are observed, the structure of the Lewis acid–acyl halide adduct is not so clearly defined. Although acyl carbonium ions are known in strongly acidic media (e.g. $RCO_2H + H^+ \rightleftharpoons RCO^+ + H_2O$, where R = 2,4,6-trimethylphenyl) the structure $RCO^+AlX_4^-$ does not seem reasonable, partly upon spectroscopic evidence and also because the exchange between $COCl_2$ and $AlCl_3$ proceeds very slowly, even though the complex is stable under reaction conditions. Perhaps a more suitable structure for the adduct involves bonding to carbonyl oxygen (6.2).

$$R-\underset{\underset{X}{|}}{C}=O\cdots\cdots AlX_3$$

(6.2)

6.5.1 *Rearrangements during Friedel–Crafts processes*

Rearrangements are frequently found during alkylation reactions. Empirically, we would expect to find *n*-alkylation rarely, because of the order of stabilities of carbonium ions,

$RCH_2^+ < R_2CH^+ < R_3C^+.$

There is no doubt that an *n*-alkyl halide will rearrange to a secondary or tertiary halide if in contact with a Lewis acid such as AlX_3; however, we must consider not only the equilibrium concentrations of these carbonium ions (where the more highly alkylated species will be the more stable) but also their relative reactivities towards the substrate in the Friedel–Crafts reaction. Under suitable conditions, it is possible to obtain alkylbenzenes such as R_2CHAr even though the predominantly formed and stable species is the tertiary carbonium ion.

Although rearrangements of the alkyl groups do not occur in Friedel–Crafts *acylation* processes, decarbonylation may be an important side-reaction. Reactions involving Me_3CCOCl can give either predominantly ketone,

MeO ⟨O⟩ + Me_3CCOCl ⟶ MeO ⟨O⟩ $COCMe_3$ (90%),

or hydrocarbon,

⟨O⟩ + Me_3CCOCl ⟶ $ArCMe_3$ + CO + HCl (90%)

and

$ArCOCMe_3$ + HCl (10%),

depending upon the reactivity of the aromatic species. Less-reactive compounds permit decarbonylation to occur

$Me_3CCOCl \rightarrow Me_3C^+ + CO + Cl^-$

and so produce alkyl-, and not acyl-benzenes.

Problems

6.1   Aromatic ethers, such as PhOMe, react readily with SO_2Cl_2 in PhCl at 25 °C to give *p*-substituted products

$ArH + SO_2Cl_2 \rightarrow ArCl + HCl + SO_2.$

The reaction is a second-order process, first order with respect to each reagent.

(a) The reaction rate is unaffected by the presence of iodine, of benzoyl peroxide, of oxygen, or of low intensity u.v. light. Is the process homolytic or heterolytic?

(b) How can it be shown that molecular chlorine formed by the dissociation of SO_2Cl_2 is not the true reagent?

(c) In $MeNO_2$ the reaction proceeds far more rapidly (by a factor of 10^5 to 10^6). Does this support your answer to (a)?

(d) In $MeNO_2$, two parallel reaction paths occur. The main one still involves attack by SO_2Cl_2, but a side-reaction involves attack by Cl_2 formed from the decomposition of the reagent:

$$\text{ArH} + \text{SO}_2\text{Cl}_2 \xrightarrow{\text{slow}} \text{ArCl} + \text{HCl} + \text{SO}_2$$

$$\text{SO}_2\text{Cl}_2 \xrightleftharpoons{\text{fast}} \text{SO}_2 + \text{Cl}_2$$

$$\text{ArH} + \text{Cl}_2 \xrightarrow{\text{slow}} \text{ArCl} + \text{HCl}$$

Derive the appropriate kinetic equations for use (i) when [ArH] and [SO_2Cl_2] are comparable, and (ii) when ArH is in great excess. Assume that the pre-equilibrium is rapidly established, but does not proceed significantly to the right.

6.2 Suggest reasons for the following:

(a) The rate of molecular chlorination of 9,10-dihydrophenanthrene in acetic acid is greater than that of biphenyl, which is in turn much greater than the rate of substitution of 2,2'-dimethylbiphenyl. (See de la Mare *et al.*, 1958.)

(b)

(See Hofmann *et al.*, 1960.)

(c) 4,6-Dibromo-1,3-dimethoxybenzene, upon treatment with nitric acid in acetic acid under nitration conditions, gives a product containing no halogen. (See Nightingale, 1947.)

(d) Chloromethylation of aromatic species (6.24) requires added Lewis acids,

$$\text{ArH} + \text{HCHO} + \text{HCl} \rightarrow \text{ArCH}_2\text{Cl} + \text{H}_2\text{O} \qquad \textbf{6.24}$$

usually ZnCl_2, as catalysts. Halogenobenzenes are less readily attacked than benzene itself, while alkyl groups accelerate the reaction. The same *ortho–para* ratio of ArCH_2X is found with alkylbenzenes regardless of the nature of the halogen acid: e.g. toluene gives the same ratio of *o*- to *p*-methylbenzyl halide whether HCl or HBr is used. (See Ogata and Okano, 1956.)

6.3 Mercuration of benzene with mercuric acetate in acetic acid is catalysed by mineral acids. The reaction is a bimolecular process, first-order with respect to both Hg(II) and ArH.

The catalysis by perchloric acid increases the selectivity of the reagent; when mixtures of benzene and toluene are used as substrate, toluene is more preferentially attacked as [HClO_4] increases. This effect stops when the perchloric acid and mercuric acetate are at equal concentrations in solution. Suggest a mechanism consistent with these results. (See Brown and McGary, 1955, Westheimer, Segel and Schramm, 1947, and Taylor and Norman, 1965, p. 194.)

Chapter 7
Aromatic Nucleophilic Substitution

1   **Activation of aromatic systems**

Simple halogenobenzenes, such as PhCl, do not readily undergo nucleophilic attack below temperatures of 200 °C. The low electron-density on carbon in the alkyl halides $(\overset{\delta+}{C}-\overset{\delta-}{X})$, due to inductive effects, is far less pronounced in the aromatic halides, since the π-electrons tend to disperse and minimize any localization of positive charge. However, electron-withdrawing groups ($-NO_2$, $-N_2^+$, $-CN$) distort the π-electron system and allow the localization of positive charge upon certain carbon atoms within the benzene ring. p-Nitrochlorobenzene, for example, reacts readily with OMe^- at 100 °C to give the corresponding ether (equation **7.1**),

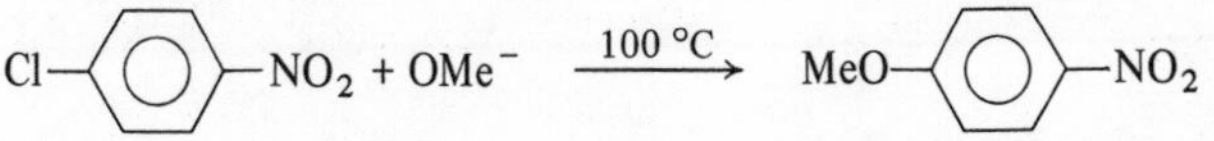

$$Cl-\!\!\!\left\langle\bigcirc\right\rangle\!\!\!-NO_2 + OMe^- \xrightarrow{100\,°C} MeO-\!\!\!\left\langle\bigcirc\right\rangle\!\!\!-NO_2 \qquad 7.1$$

while similar displacement of chlorine from 2,4-dinitrochlorobenzene occurs at room temperature, and 2,4,6-trinitrochlorobenzene reacts readily with much weaker nucleophiles. Similarly, rearrangements can occur during diazotization in solutions of hydrogen halides; 2,6-dichloroaniline in 48 per cent HBr gives, upon diazotization and decomposition, products in which one or both chlorine atoms have been replaced by bromine **(7.2).**

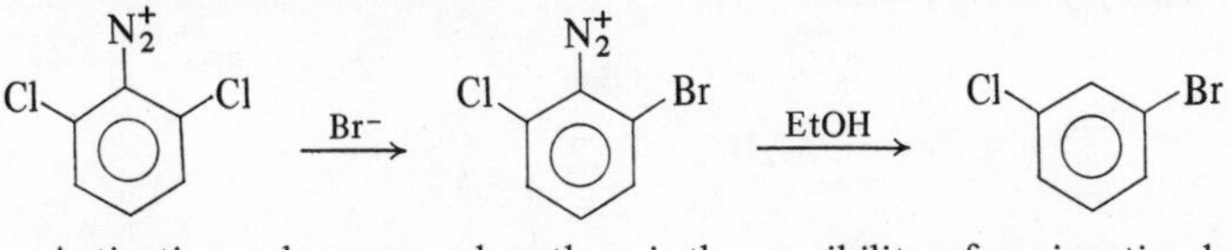

$$\overset{N_2^+}{Cl\!\!\underset{\bigcirc}{\diagup}\!\!Cl} \xrightarrow{Br^-} \overset{N_2^+}{Cl\!\!\underset{\bigcirc}{\diagup}\!\!Br} \xrightarrow{EtOH} Cl\!\!\underset{\bigcirc}{\diagup}\!\!Br \qquad 7.2$$

   Activation only occurs when there is the possibility of conjugation between the substituent and the α-carbon atom; m-chloronitrobenzene is almost as inert to nucleophilic attack as chlorobenzene itself.

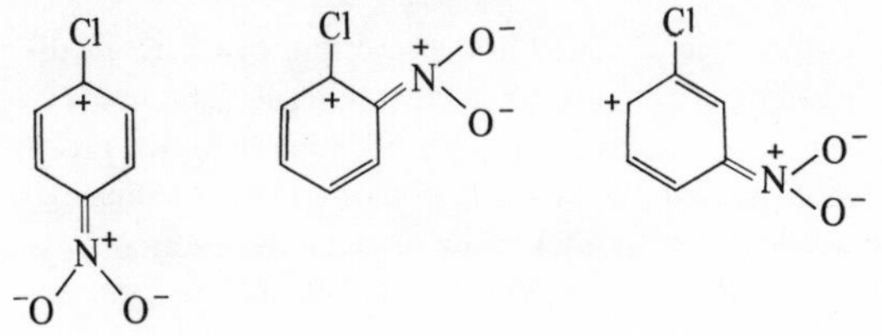

7.2   Bimolecular nucleophilic substitution

Nitrohalogenobenzenes have been studied in reaction with a large number of nucleophiles. The processes almost always involve second-order kinetics,

Rate $= k[\mathrm{ArX}][\mathrm{Y}^-]$,

and have been shown to be nucleophilic substitutions since electron-withdrawing substituents in the aryl halide facilitate the reaction (Hammett's ρ for 2-nitro-4-X-chlorobenzenes reacting with OMe^- in MeOH = 3·9). The detailed mechanism of the displacement need not be the same as for aliphatic substitution. The C—X bond breaking process cannot be involved in the rate-determining stage, for the order of reactivity of 2,4-dinitro-X-benzenes with OMe^- in MeOH is $F \gg Cl > Br > I$, the exact reverse of the ease of reaction of alkyl halides, where partial bond breaking occurs in the transition state, and the order in which the C—X dipole is developed. Similarly, the rates of reaction of piperidine with some 2,4-dinitro-X-benzenes (X = Cl, Br, SOPh, and $\mathrm{SO_2Ph}$) are almost identical, as are the activation energies for the four processes. Apparently the slow stage of the reaction involves attack of the nucleophile upon the aromatic carbon atom before bond-breaking processes are well developed (equation 7.3).

7.3

Whether Wheland-type intermediates exist, or whether they merely represent an activated complex of no inherent stability, is still a source of controversy. Meisenheimer (1882) prepared a species, thought to be such an intermediate, either by reacting KOMe with 2,4,6-trinitrophenetole, or by reacting KOEt with 2,4,6-trinitroanisole; upon decomposition with acid both samples have the same mixture of methyl and ethyl ethers (7.4).

7.4

No similarly stable species have been isolated in the reactions of mono- or di-nitrobenzene derivatives, but there is evidence for such an intermediate in the substitution of 2,4-dinitrofluorobenzene by N-methylaniline in ethanol, a process which is catalysed by potassium acetate. If the formation of a Meisenheimer complex is reversible, base would allow another mode of decomposition and would therefore increase the overall rate of reaction.

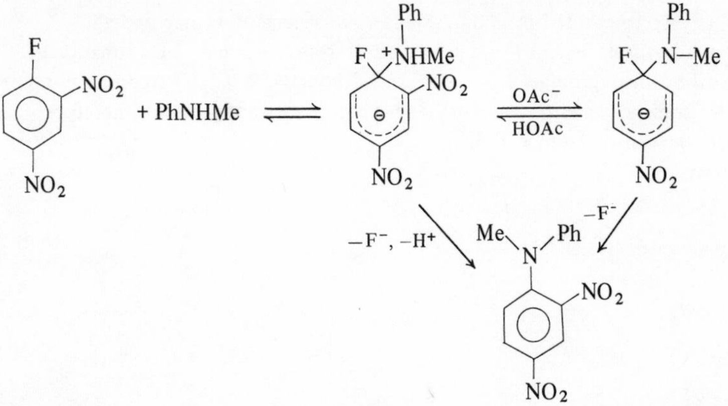

Such catalysis is not found with the corresponding chloro- or bromo-dinitrobenzenes, presumably because of the different free energies of the component reactions.

2.1 *Steric effects in bimolecular nucleophilic substitution*

Activation by a nitro group requires conjugation between the substituent and the reaction site. This in turn requires that the $N\overset{+}{\underset{O^-}{\diagup}}\!\!\!\!\overset{O}{}$ group becomes nearly coplanar with the benzene ring, and if this is prevented, by bulky *ortho*-substituents, the susceptibility of an aromatic halide to bimolecular attack will be drastically reduced. For instance, 2-bromo-5-nitro-*m*-xylene (7.1) reacts much less readily than *p*-bromonitrobenzene with piperidine.

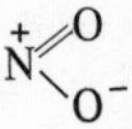

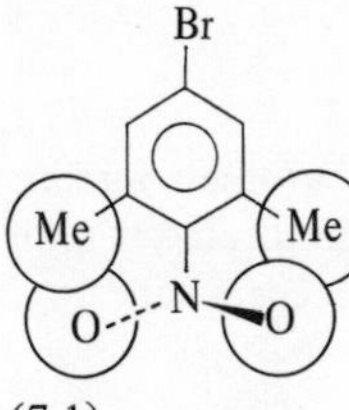

(7.1)

because the activating nitro-group cannot attain coplanarity. Similarly, 1-nitro-2-halogenonaphthalenes (7.2) react less readily ($\delta \Delta E \simeq 8$ kJ mol^{-1}) than the 2-nitro-1-halogenonaphthalenes (7.3) since the 1-nitro group suffers steric hindrance from hydrogen in the 8- (peri-) position (7.2)

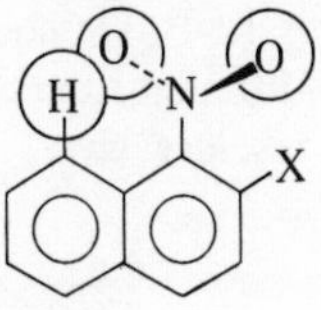

(7.2) 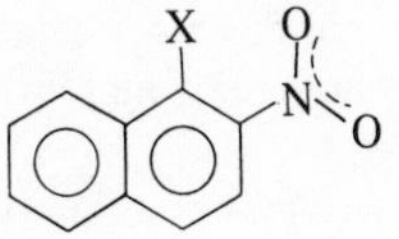   (7.3)

In principal, either nitro group in 2,5-dinitro-m-xylene (7.4) may undergo nucleophilic displacement through activation from the other $-NO_2$ substituent; in fact the 2-nitro group is displaced, since although it is subject to primary steric effects for displacement, it cannot achieve coplanarity and may not therefore activate the carbon atom at C-5.

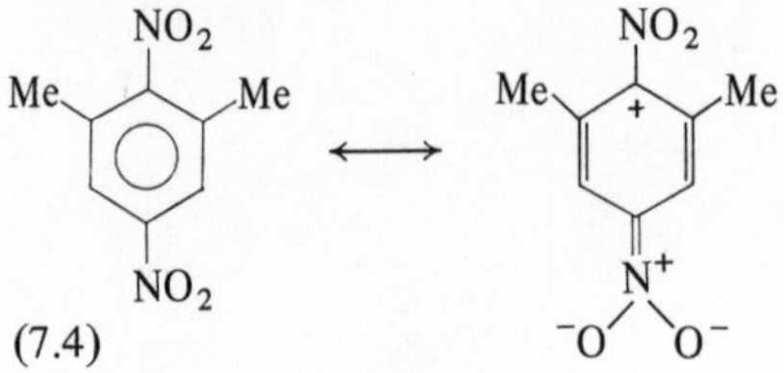

(7.4)

7.3 Unimolecular nucleophilic substitution

The analogue of the aliphatic S_N1 mechanism would require an aromatic system in which the breakage of the C—X bond can be the rate-determining process. This has not been realized with aromatic halides, even under the most suitable conditions (solvolysis of 2,4,6-trinitrochlorobenzene). However, the decomposition of aromatic diazonium ions in water are kinetically first-order processes; although the *rate* of reaction (apart from solvent and salt effects) is not altered upon addition of methanol or of chloride ion, the *products* of decomposition now include PhOMe and PhCl respectively. A similar situation occurs in the aliphatic S_N1 processes, and it is suggested that the slow stage of the diazonium reactions involves heterolysis of the Ar—N bond (equations 7.5, 7.6).

$$ArN_2^+ \xrightarrow{\text{slow}} Ar^+ + N_2 \qquad\qquad \textbf{7.5}$$

$$Ar^+ + X^- \xrightarrow{\text{fast}} ArX \qquad\qquad \textbf{7.6}$$

The rates of decomposition of a number of substituted phenyl diazonium ions in water at 29 °C give a Hammett plot with $\rho \simeq -4$, in keeping with this mechanism. With more potent nucleophiles, the reaction is not purely S_N1 in character,

$$\text{Rate} = k[ArN_2^+] + k_2[ArN_2^+][Br^-],$$

a situation also found in aliphatic chemistry (cf. benzyl halides).

7.4 Arynes

The inertness of aromatic halides (in the absence of activating groups) to nucleophilic attack is in sharp contrast with their reaction with strong bases (e.g. KNH_2) in liquid ammonia at -35 °C. The mechanism of this process must be very different from the bimolecular substitutions already considered. Chlorobenzene, for instance, gives aniline

$$PhCl \xrightarrow[\text{NH}_3]{\text{NaNH}_2} PhNH_2,$$

but if the carbon atom bearing the halogen is labelled, only 48 per cent of the aniline produced has the amino group bonded to the labelled carbon (equation **7.7**).

48%   52%   **7.7**

Similarly, *o*-chlorotoluene gives a mixture of *o*- and *m*-toluidine; *m*-chlorotoluene produces all three toluidines; and *p*-chlorotoluene provides a mixture of *m*- and *p*-toluidines. At some stage in the reaction two adjacent carbon atoms in benzene appear to become equivalent, or at least both can be attacked by NH_2^- to give products. The most direct explanation of this is the postulation of benzyne, or an aryne (equation **7.8**).

7.8

The formulation of an aryne intermediate also explains the isolation of a mixture of both α- *and* β-naphthyl piperidines in the *same* proportions, from treating α-chloro-, bromo-, or iodo-naphthalene with piperidine and sodium amide. As 68 per cent of the product is the α-isomer, naphthyne is not symmetrical (the two carbon atoms involved are not attacked with equal facility). Various structures for benzyne have been proposed; all have disadvantages. Benzyne does not contain a triple bond, for the strain in the ring system would surely be prohibitive; nor can it be regarded as a zwitterion, since attack can occur at either carbon atom.

Benzyne also shows considerable dienophile reactivity in Diels–Alder syntheses. Its intermediacy in the reaction of *o*-fluorobromobenzene with lithium or with magnesium is shown (i) by the isolation of the Diels–Alder adduct when furan is present in the reaction mixture, and (ii) the formation of tryptacene when anthracene is present in the reaction mixture

Also small yields of biphenylene and triphenylene are found, probably from dimerization and trimerization of the benzyne radicals.

7.4.1 Requirements for the aryne mechanism

Since the elimination–addition mechanism (benzyne mechanism) requires firstly the removal of HX from a halogenobenzene, such processes can only occur in the presence of strong base, and where there is at least one *ortho*-hydrogen atom available. Significantly, bromomesitylene (2,4,6-trimethylbromobenzene), bromodurene (3-bromo-1,2,4,5-tetramethylbenzene) and 2-iodo-*m*-xylene are inert to sodium amide in liquid ammonia under conditions where less-substituted halogenobenzenes react. However, the presence of strong base and an available proton does not automatically ensure the benzyne mechanism. The base-catalysed hydrolysis of halogenobenzenes and halogenotoluenes at 250–350 °C can occur either by the benzyne process, or by a simple S_N2 process, or by a combination of both (shown by the extent of rearrangement). Similarly the reaction of α-fluoronaphthalene with piperidine and sodium amide gives more than 68 per cent α-naphthylpiperidine (cf. section 7.4) implying the operation of a direct S_N2 displacement in addition to the aryne mechanism.

Problems

7.1 In an attempt to show that C—Cl bond-breaking is not significant in the reaction between 2,4-dinitrochlorobenzene and OMe⁻ in MeOH, the rate of reaction of dinitrochlorobenzene containing only ³⁶Cl was compared with that of the substrate containing chlorine in the natural isotopic abundance (³⁵Cl : ³⁷Cl = 3 : 1).
   Is this experiment likely to give significant results?

7.2 Although fluorobenzene does not readily react with NH_3, or with PhS⁻, at temperatures below 250 °C, hexafluorobenzene reacts readily with these nucleophiles (in ethanol solution) at room temperature. Explain this observation.

7.3 *p*-Nitroacetanilide in strongly basic solution (OH⁻ in aqueous EtOH) gives *p*-nitrophenol on prolonged heating. How could the reaction be followed? How could one differentiate between a mechanism involving direct displacement of CH_3CONH^- by OH⁻, and one in which the acetamido group was hydrolysed, and the amino group subsequently displaced by OH⁻? Construct possible mechanisms and design experiments to verify or disprove them.

7.4 (a) *N*-Methyl-3-(*m*-chlorophenyl)propylamine on treatment with ethereal solutions of phenyl lithium gives an *N*-methyl-tetrahydroquinoline. Explain the course of the reaction.

(b) Under strongly basic conditions (OEt⁻ in EtOH) hydroxylamine reacts with 1-nitronaphthalene to give 4-nitro-1-naphthylamine. Suggest a mechanism for the reaction, remembering that both base and the nitro group seem to be essential.

Chapter 8
Reactions at Unsaturated Linkages

8.1   **Introduction**

One of the most characteristic reactions of olefins is addition to the multiple bond; both the qualitative ($KMnO_4$, Br_2) and quantitative (ICl – 'iodine numbers') detection of olefins rely upon such processes (e.g. equation **8.1**).

$$RCH{=}CH_2 + Br_2 \rightarrow RCHBrCH_2Br \qquad\qquad\qquad\qquad 8.1$$

In fact, addition reactions can be observed with all multiple bonds, and we will consider not only reactions involving olefins and acetylenes, but also those in which carbon–oxygen, carbon–nitrogen, and nitrogen–nitrogen bonds are attacked.

In theory, both nucleophilic and electrophilic addition processes are possible; in practice, electrophilic addition reactions are more important in olefin chemistry, and nucleophilic reactions occur more easily in acetylene chemistry.

8.2   **Electrophilic additions to carbon–carbon multiple bonds**

8.2.1   *Acid-catalysed hydration*

The formation of alcohols from olefins occurs readily in dilute acid, e.g.

$$RCH{=}CH_2 + H_2O \xrightarrow{\;H^+\;} RCH(OH)CH_3.$$

The orientation of addition (Markovnikov rule) is determined by the relative stabilities of the two possible carbonium ions which are intermediates. In the present example, $RCH_2CH_2^+$ is less stable than $R\overset{+}{C}HCH_3$, which is therefore the precursor of the reaction product. Having implied an electrophilic addition mechanism, let us examine the evidence.

The hydrations of olefins in the same acid solutions accelerate in the order

$$CH_2{=}CH_2 < RCH{=}CH_2 < RCH{=}CHR < R_2C{=}CHR < R_2C{=}CR_2.$$

More generally, the rate of hydration is increased by alkyl or phenyl substituents attached to the olefin system, and decreased when electron-withdrawing substituents ($-NO_2$, $-CO_2H$, $-CHO$, etc.) are attached. In dilute acid solution, the reaction follows the kinetic form

$$\text{Rate} = k[H^+]\,[\text{olefin}],$$

but in more concentrated acidic media, the rate depends upon h_0 (section 2.9).[*]
This implies that the slow stage involves the protonated system

$$\text{>C=C<} \;\underset{}{\overset{H^+}{\rightleftharpoons}}\; \text{>CH–\overset{+}{C}<}$$

(cf. B $\overset{H^+}{\rightleftharpoons}$ BH$^+$),

whose rate of formation, or of subsequent reaction, determines the overall rate of
hydration. Until recently, the reaction was believed to be subject to *specific* acid
catalysis (section 2.7), but the hydration of *p*-methoxystyrene in formic acid–
formate buffers involves catalysis both by H$^+$ and by HCO$_2$H, and so is subject to
general acid catalysis.

The nature of the intermediate is not settled. While π-complexes almost certainly
exist between olefins and H$^+$ (as they do with many Lewis acids, e.g. Ag$^+$, halogens)
they seem unlikely to correspond to the transition-state configuration, for such
complexes are usually formed readily and reversibly. General opinion favours a
carbonium ionic intermediate, but there is still discussion on whether the ion is
'encumbered' (by solvent molecules held in a particular orientation) or unencumbered,
and to what extent the ionic character of the intermediate is developed (8.1–3).

π-complex	encumbered carbonium ion	unencumbered carbonium ion
(8.1)	(8.2)	(8.3)

A mechanism such as

reflects both the observations in the hydration of olefins and other experimental
work on the dehydration of alcohols (reverse process).

The hydration of α,β-unsaturated carbonyl compounds (e.g. MeCH=CHCHO)
appears to depend upon [H$^+$] rather than upon h_0, or to show an acid-dependence
intermediate between these two functions. Mechanisms for such processes have
been proposed, usually involving nucleophilic attack upon the conjugate acid of the
carbonyl derivative. These systems, which are fairly heavily deactivated towards
electrophilic attack upon the C=C system, might be expected to undergo reaction
by some more easy, nucleophilic process (cf. halogen addition).

---

[*] The departure from the theoretical prediction (Rate $\propto h_0^{1.00}$) is explained by considering the
activity coefficient of the intermediate; if allowance is made for these effects, a dependence
much closer to the theoretical value is found.

The hydration of acetylenes occurs in acid media. The mechanism of the process

$$RC\equiv CH + H_2O \xrightarrow{\ H^+\ } [RC(OH)_2CH_3] \longrightarrow RCOCH_3$$

is not certain, for there have been relatively few investigations of this system, partly because of the higher acidity required (bringing in complications of the relevant acidity function, see section 2.9.4) and partly because of the mercury-catalysed process which takes place more readily.

2.2   *Addition of* HX

Both strong acids (e.g. HCl, HBr) and weak acids (e.g. MeOH) will add across olefinic bonds (equation 8.2).

$$RCH=CH_2 + HX \rightarrow RCHXCH_3 \qquad\qquad 8.2$$

The effects of alkyl and aryl substituents upon the rates of these additions make it clear that the slow stage is an electrophilic attack by H^+ or some kinetically equivalent species. In non-polar solvents, the kinetic order with respect to HX is high (ranging between two and four), and the rate of the reaction is very sensitive to adventitious catalysts such as water. At least some of the molecules of HX appear to fulfil the role of a solvent, although the possibility of species such as H_2Cl_2 (which may be better proton-donors) cannot be excluded.

In polar solvents the mechanism of these additions seems to be similar to that proposed for acid-catalysed hydration of olefins, with X^- or, in the case of weak acids, ROH taking the place of water as the nucleophile. Addition takes place almost entirely in the *trans* sense,

and the similar, but slower, addition to triple bonds also provides predominantly *trans*-adducts,

2.3   *Addition of halogens*

In non-polar media, the reaction of chlorine or bromine with olefins proceeds erratically. Attempts to determine the order of the reaction with respect to either reagent are complicated, not only by the magnitude of these kinetic orders (e.g. in iodine addition, rate $\propto [I_2]^3$) but by adventitious catalysis, not only by trace impurities (e.g. water, alcohol) but also by free-radical sources (e.g. u.v. light, oxygen) which initiate a homolytic reaction path (section 11.4.2).

05   Electrophilic additions to carbon–carbon multiple bonds

In acetic acid, or more polar solvents (e.g. nitromethane), a heterolytic reaction mechanism has been clearly defined. We have already mentioned (section 3.2.2) the evidence for a carbonium ionic intermediate in the addition of bromine to ethylene in aqueous media; substituent effects clearly indicate that halogen addition involves electrophilic attack in the rate-determining stage, providing additional evidence of such an intermediate.

$$>C=C< \ + X_2 \longrightarrow -\overset{|}{\underset{+}{C}}-\overset{|}{\underset{|}{C}}- \qquad\qquad 8.3$$
$$\underset{X}{}$$

Since the concentration of nucleophile in the solution (after complexing effects have been allowed for, e.g. the formation of Br_3^- from Br^-) does not seem to affect the rate of bromination of olefins, we conclude that the first stage of the addition (equation 8.3) and not the attack by the nucleophile upon the carbonium ionic species is rate-determining.

Chlorine additions usually show second-order kinetics,

Rate = $k[Cl_2]$ [olefin],

and similar kinetic forms are shown by bromine and by some of the interhalogens (e.g. ICl, IBr, BrCl) in acetic acid or nitrobenzene at low concentrations of reagent. A higher-order term can be distinguished, and operates exclusively at concentrations of about 0·1 M and above under these conditions (8.4).

Rate = $k[X_2]$ [olefin] + $k_3[X_2]^2$ [olefin] $\qquad\qquad$ 8.4

A number of explanations have been offered; the most probable seems to be that in such solvent media the formation of the carbonium ionic intermediate is assisted by a molecule of halogen or of interhalogen. The reaction sequence

$$>C=C< \ + X_2 \ \underset{\text{}}{\overset{\text{fast}}{\rightleftharpoons}} \ \underset{\underset{\underset{\delta+ \quad \delta-}{X\cdots X}}{\downarrow}}{>C=C<} \ \overset{\text{slow}}{\longrightarrow} \ >\overset{|}{\underset{+}{C}}-\overset{|}{\underset{X}{C}}\diagdown \ + X^-$$

implies that an effect which will assist breaking the X—X bond, or stabilize $X^{\delta-}$, should accelerate the reaction. The solvent may be only weakly able to solvate X^-, but the known stability of Br_3^- suggests that halogen might take over the solvent function and thereby assist the reaction. At low concentrations, halogen interaction with $X^{\delta-}$ can only be small (hence second-order kinetics), but such interaction can dominate the reaction course at higher concentrations. The low stability of Cl_3^- is in keeping with the second-order kinetics observed for chlorine addition at all concentrations.

Again, the structure of the intermediate carbonium ionic species is controversial. Since halogen addition occurs in an almost exclusively *trans* sense, chloronium (8.4) or bromonium (8.5) ions

$$\underset{\text{(8.4)}}{\overset{\displaystyle \ge C\!\!-\!\!-\!\!C\le}{\underset{\overset{\displaystyle +}{\ddot{C}l}}{}}} \quad \text{or} \quad \underset{\text{(8.5)}}{\overset{\displaystyle \ge C\!\!-\!\!-\!\!C\le}{\underset{\overset{\displaystyle +}{Br}}{}}}$$

have been suggested as intermediates. As in the case of the α-lactone in neighbouring-group participation by $-CO_2^-$ (section 4.3.2), the bond between halogen and C_β need not be fully developed to explain the stereochemistry of the process. Some interaction certainly exists; in the reaction of HOCl with $CH_2{=}CHCH_2{}^{36}Cl$, some of the 2,3-dichloropropanol formed has radioactive chlorine attached to the *second* carbon atom, implying a rearrangement involving an ion such as (8.6), in which

$$\underset{\text{(8.6)}}{\overset{{}^{36}Cl}{H_2C\!\!-\!\!\underset{\underset{Cl}{H}}{\overset{+}{C}}\!\!-\!\!CH_2}}$$

both the entering halogen and the radioactive isotope can interact with the carbonium ionic centre. Interactions between halogen atoms and adjacent carbonium ionic centres are well known (from solvolysis reactions and from rearrangements) but the precise degree of bonding which exists is still uncertain.

3 **Nucleophilic additions to carbon–carbon multiple bonds**

The addition of halogen to α,β-unsaturated aldehydes in acetic acid proceeds very slowly initially, but is then subject to vigorous autocatalysis. This catalysis can be eliminated by the presence of sodium acetate, and can be induced by the addition of acids. The effect is not purely the result of the proton, for HCl and HBr are far more effective catalysts (although much weaker acids in acetic acid solution) than $HClO_4$ or H_2SO_4. In the absence of added species, the autocatalysis seems to arise from HX formed by side-reactions. The most likely explanation seems to involve the protonation of carbonyl oxygen, followed by *nucleophilic* attack upon the allylic carbonium ion,

$$\ge C{=}CC{=}O \;\overset{H^+}{\rightleftharpoons}\; \underset{+}{\underbrace{\ge C\text{---}C\text{---}C}}{-}OH \;\overset{Br^-}{\longrightarrow}\; \underset{\overset{|}{Br}}{-}C{-}C{=}C{-}OH \;\overset{Br_2}{\longrightarrow}$$

$$\underset{\overset{|}{Br}}{-}\overset{\overset{|}{Br}}{C}{-}\overset{\overset{|}{}}{C}{-}\overset{+}{C}{-}OH.$$

While this mechanism is consistent with the facts, there is insufficient evidence to rule out possible alternatives. Other 'deactivated' (towards electrophilic attack) systems also undergo nucleophilic addition reactions, although usually with very potent nucleophiles such as carbanions (as in the Michael condensation, Chapter 12) or thiols (RS^-).

The acetylenes, which we have seen to be less susceptible to electrophilic attack than the olefins, more usually undergo nucleophilic addition processes. The addition of HI (from lithium iodide in acetic acid solvent) to C≡C is a bimolecular process whose rate increases in the order

$$RC \equiv CR < RC \equiv CCO_2Me < MeO_2CC \equiv CCO_2Me$$

and similar evidence of nucleophilic addition mechanisms are found in the reaction of azides, and of thiols, with acetylenes.

8.4   **Additions to C=O, C=N and N=N**

Solutions of formaldehyde in water seem to contain predominantly methylene glycol ($CH_2(OH)_2$), although this can be easily dehydrated, so that formaldehyde can be distilled from such solutions. In general, *stable* hydrates are formed (e.g. chloral hydrate) only when several electron-withdrawing groups are present. Interestingly, the carbonyl group itself can function as such a stabilizing entity, for 1,3-diphenyl-1,2,3-propantrione (PhCOCOCOPh) crystallizes from water as the hydrate $PhCOC(OH)_2COPh$. Many other reactions, including the formation of the condensation products (oximes, hydrazones, etc.) usually used to characterize carbonyl derivatives, involve addition across C=O initially.

The C=O group will be strongly polarized, and electrophilic attack at carbon would seem to be easily accomplished. The hydration of acetaldehyde, and of some other simple carbonyl derivatives, is subject to general acid catalysis, as is the dehydration process. The corresponding reaction with methanol (forming a hemi-acetal) also involves general acid catalysis,

RCHO + MeOH → RCH(OMe)OH.

A plausible mechanism incorporating these features is

$$RCHO + HX \rightleftharpoons \underset{\substack{\text{(or the fully} \\ \text{formed conjugate} \\ \text{acid)}}}{RCHO\cdots HX} \xrightleftharpoons{\text{H}_2\text{O, slow}} \underset{\substack{| \\ \overset{+}{O}H_2 \\ \downarrow}}{RCH-O^-} \; HX$$

$$RCH(OH)_2 + HX \longleftarrow RCH(OH)_2\cdots HX$$

The initial steps in the process have some similarity to the formation of alcohols by acid-catalysed hydration of olefins. A dependence upon h_0 might be expected, and has been observed in the hydration of some aldehydes (see, however, section 8.2).

$$RCOR' + H_2NX \rightarrow [RR'C(OH)NHX] \rightarrow RR'C=NX \qquad 8.5$$

The formation of oximes, phenylhydrazones and semi-carbazones also show general acid catalysis. Here again we could imagine attack by the unshared pair of electrons on nitrogen upon a complex, or conjugate acid (RCOR' $\cdots$ HX, or $RR'\overset{+}{C}(OH)X^-$), of the carbonyl compound, as the rate-determining step. The detection of acid catalysis is not easy, since increasing the acidity of the solution also decreases the effective concentration of the reagent, which is converted to the unreactive conjugate acid.

The formation of oximes is also catalysed by strong base, probably due to the formation of the more reactive nucleophile $NHOH^-$ (or $H_2N{-}O^-$) which may react rapidly with the positively charged carbon of the $C{=}O$ system without its further activation through protonation. The weaker acids, semicarbazide and phenylhydrazine, have not been reported to show base catalysis.

Many other reactions, such as the aldol condensation and the Claisen condensation, involve nucleophilic attack upon $C{=}O$ or $C{=}C$ systems. They shall be considered in Chapter 12.

3.4.1 N=N *systems and related systems*

The addition of diazonium cyanides ($ArN{=}NCN$) to 9-diazofluorene illustrates addition across $N{=}N$. The process is reported to involve electrophilic attack by ArN_2^+, but the process may equally as well be initiated by the strong nucleophile, CN^- (equation **8.6**).

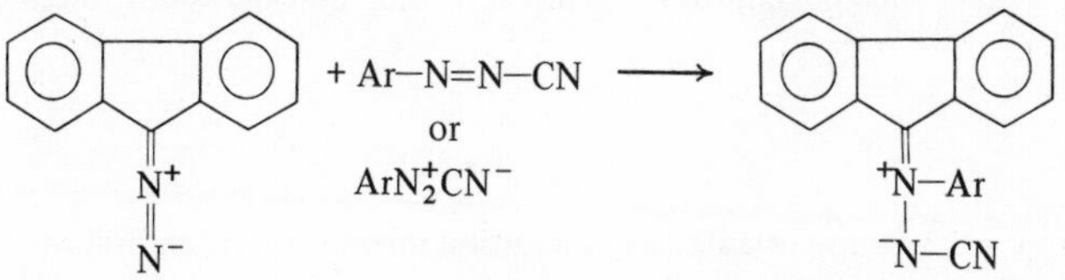

$$\text{+ Ar–N=N–CN} \quad\text{or}\quad ArN_2^+CN^- \longrightarrow \qquad\qquad\qquad \textbf{8.6}$$

The reaction of bromine with azobenzene ($PhN{=}NPh$) is catalysed by hydrogen bromide. This is consistent with an initial electrophilic addition of HBr across the $N{=}N$ bond *covalently* (a salt would deactivate the electrophilic substitution process) so that one aromatic ring is activated by an electron-rich nitrogen atom:

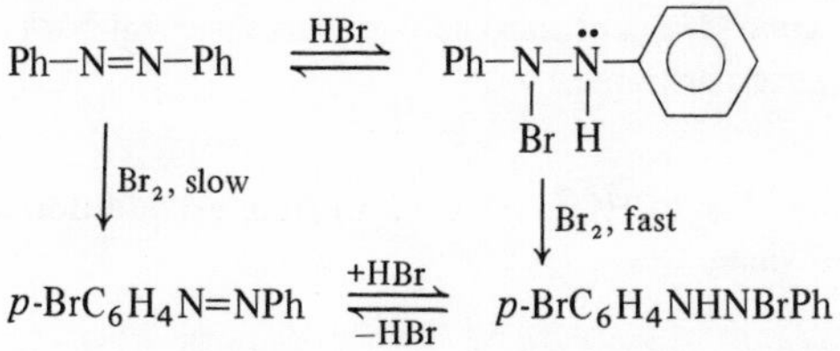

Similar evidence is found for the addition of HBr across $N{=}O$ in nitrosobenzene; again, hydrogen bromide catalyses the electrophilic attack by bromine upon the ring.

Addition of acids to $C{=}N$ in Schiff's bases is a general reaction, although there is controversy over the nature of the product. Evidence has been found for both the ionic form ($RCH\overset{+}{N}HR'X^-$) and for the covalently bound adduct ($RCHXNHR'$); the latter species, on treatment with water, appears to give a hydration product, whereas the former reverts to the Schiff's base under these conditions. Such additions are responsible for the hydration of a number of heterocyclic compounds, where the heterocyclic ring formally contains the $C{=}N$ system and so reacts as a Schiff's base.

Pteridine, on treatment with dilute acid, forms 3,4-dihydro-4-hydroxypteridine (equation **8.7**) and the reaction seems to be general for a number of polyaza-

$$8.7$$

naphthalenes. The rate of the reaction is strongly affected by the acidity of the medium, for not only is the hydrolysis reaction of the heterocycle subject to general acid catalysis and base-catalysis, but in some cases two 'hydrates' are formed. One is formed and decomposed rapidly, and is so subject to kinetic control; the other is formed slowly but irreversibly (under the reaction conditions) so is subject to thermodynamic control. The difference in these two conditions is that the kinetically controlled product is the one formed most *rapidly*, assuming that the reverse reaction has not become significant; the thermodynamically controlled product is the one which is least prone to reverting to the starting materials (i.e. that for which the rate of the *reverse* reaction is very slow).

In consideration of such possibilities, it is not surprising that the kinetic picture is quite complex.

Problems

8.1 An early attempt to show the orientation of addition involved addition, followed by elimination. Since it was thought that elimination proceeded in the *trans*-diaxial sense, the direction of addition could be shown by considering the stereochemistry of the starting and finishing materials. Show (i) that the addition of halogen and elimination of HX from either a *trans* or a *cis* olefin gives a product of the same stereochemistry (if addition proceeds *cis*) or of opposite stereochemistry (if addition proceeds *trans*) and (ii) that the *same* stereochemical consequences occur if *both* elimination and addition are *either* cis *or* trans.
(See Auwers, 1923.)

8.2 Suggest a reason for the isolation of $Me_3CCHClCHMeCMe_2Cl$ from the addition of chlorine to *trans*-1,2-di-*t*-butylethylene.
(See Puterbaugh and Newman, 1959.)

8.3 2,4-Dinitrobenzenesulphenyl chloride reacts with styrene by a second-order process. In acetic acid at 25 °C, the relative rates of reaction of some substituted styrenes are p-MeO, 325; p-Me, 3·1; H, 1·0; p-Cl, 0·3; p-NO$_2$, 0·019.
How could you show (i) that the process is an electrophilic addition, and (ii) that the attacking species involves $ArS^{\delta+}$ and not $Cl^{\delta+}$?

8.4 Suggest for the following observations:

(a) The rate of addition of HCl to a number of olefins in benzene is of high order

$$\text{Rate} = k[\text{olefin}] [\text{HCl}]^n \quad (n \leqslant 3)$$

but the reaction is accelerated, and the order with respect to HCl is diminished, when $SnCl_4$ is added.

(b) Crotonaldehyde reacts with Cl_2 in HOAc *less* rapidly than crotonic acid in the presence of sodium acetate, but *more* rapidly than crotonic acid in the presence of HCl or $HClO_4$.

(c) At low temperatures, and using a deficit of HCl, isoprene adds HCl to provide $Me_2C(Cl)CH=CH_2$. If an excess of reagent is used, the product is $Me_2C=CHCH_2Cl$.

8.5 The formation of epoxides from olefins and per-acids,

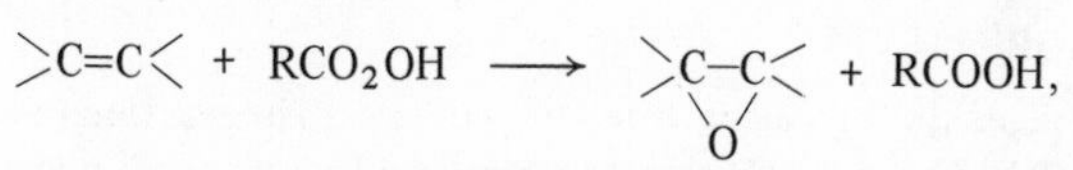

$$\text{>C=C<} + RCO_2OH \longrightarrow \text{>C—C<} + RCOOH,$$
$$\overset{\diagdown}{O}\diagup$$

is accelerated by the presence of alkyl groups in the olefin, and is retarded by aldehyde and carboxylate groups attached to the double bond. The attack by substituted perbenzoic acids upon stilbene in benzene solution is facilitated by p-NO_2 and p-Cl, but retarded by p-OMe groups. However, the rate of the epoxidation is not greatly altered by changes in the solvent polarity.

(a) Suggest a mechanism to account for these observations.

(b) Show how the absence of isotopic exchange between perbenzoic acid and benzoate ion implies the non-existence of OH^+ in these solutions.

(c) From this, show how the reaction can be subject to catalysis by trifluoroacetic acid.

8.6 Complete the following reactions:

(a) 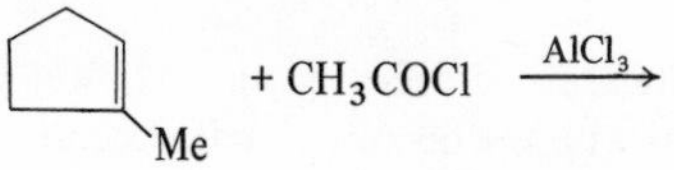 $+ CH_3COCl \xrightarrow{AlCl_3}$

(b) $MeCH=CH_2 + CF_3CO_2H \xrightarrow{MeOH}$

(c) $CH_2=CHOCH_3 \xrightarrow{H^+/H_2O}$

Chapter 9
Reactions of Carboxylic Acid Derivatives

9.1   Nomenclature in ester hydrolysis

The formation of an ester,

$$R'OH + RCO_2H \rightleftharpoons RCOOR' + H_2O,$$

could, in principle, take place through eight different processes. The reaction (and the reverse hydrolysis) might be a unimolecular or a bimolecular process; it might involve (in the case of ester hydrolysis) either the unprotonated ester or its conjugate acid ($RCO\overset{+}{O}HR'$), and it might proceed either by the formation (or fission) of an acyl-oxygen bond (RCO—O) or of an alkyl-oxygen bond (RCOO—R').

The molecularity of the reaction can be designated by 1 or 2, as in the S_N mechanisms. Whether the true reagent is the species itself or the conjugate acid can be indicated by B or by A, respectively, and whether the process involves acyl-oxygen or alkyl-oxygen fission can be shown by the subscripts AC or AL. This is a convenient shorthand form used to describe esterification reactions; thus, $A_{AL}1$ would be a *unimolecular* process involving the conjugate *acid* of the species and *alkyl*-oxygen fission.

9.2   Saponification – the $B_{AC}2$ mechanism

Base-catalysed hydrolyses of esters are often second-order reactions. In water containing ^{18}O (which rapidly equilibrates with the dissolved hydroxide ion

$$OH^- + H_2{}^{18}O \rightleftharpoons H_2O + {}^{18}OH^-),$$

the oxygen isotope appears in the carboxylate ion only (equation **9.1**).

$$RCOOR' + {}^{18}OH^- \rightarrow [RCO^{18}O]^- + R'OH \qquad \textbf{9.1}$$

The rate of saponification increases when electron-attracting substituents are attached to alkyl or acyl groups of the ester, and hence involves nucleophilic attack (by OH^-) upon the substrate.

A mechanism consistent with these results is

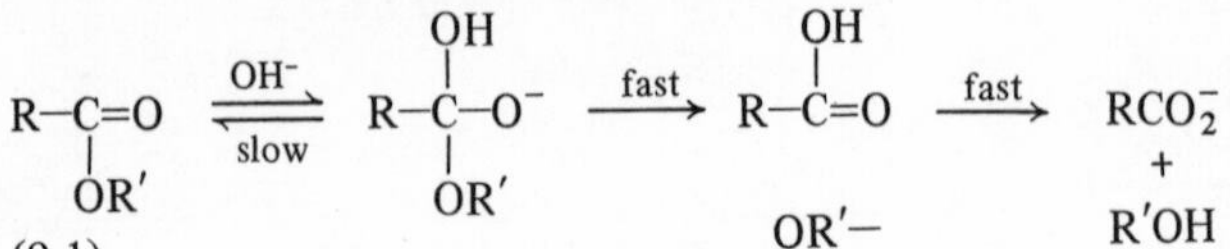

(9.1)

In this way, ^{18}O from hydroxide ion becomes incorporated solely in the carboxylate anion product, the slow stage of the reaction is a bimolecular process, and the substituent effect is explained.

The intermediate ion (9.1) does not correspond to the activated-complex configuration. If the saponification of an ester with ^{18}O at carbonyl oxygen $(RC^{18}OOR')$ is interrupted, the residual ester is found to have lost some of the isotope. This can best be explained by exchange between (9.1) and the solvent (equation 9.2).

$$\left[\begin{array}{c} OH \\ | \\ R-C-^{18}O \\ | \\ OR' \end{array}\right]^{-} + H_2O \rightarrow \begin{array}{c} O^- \\ | \\ R-C-^{18}OH \\ | \\ OR' \end{array} \rightarrow \begin{array}{c} O \\ \| \\ R-C \\ | \\ OR' \end{array} + ^{18}OH^- \qquad 9.2$$

Since (9.1) must exist for an appreciable time for such exchange to be detectable, it cannot be an activated complex (section 2.3).

9.3 Acid-catalysed hydrolysis

9.3.1 $A_{AC}1$ *mechanism*

2,4,6-Trimethylbenzoic acid is resistant to esterification by 'normal' methods (MeOH–HCl, MeOH–H_2SO_4), and the ester, once formed, is similarly resistant to hydrolysis. This probably reflects the steric effect of the *ortho*-substituents, which hinder the approach of nucleophiles needed to allow either esterification or hydrolysis to occur. In contrast, the solution of trimethylbenzoic acid in sulphuric acid gives methyl 2,4,6-trimethylbenzoate when poured into cold methanol; the sulphuric acid solution of the ester, when poured into cold water, forms the acid. This is rationalized by the intermediacy of the acyl carbonium ion (equation 9.3) whose formation involves protonation at a less-hindered atom

$$ArCOOH \xrightarrow{H_2SO_4} ArCO^+ \xleftarrow{H_2SO_4} ArCOOMe \qquad 9.3$$

$$ArCOOMe + H^+ \overset{MeOH}{\nearrow} \qquad \overset{H_2O}{\searrow} ArCOOH + H^+$$

and also allows considerable relief of steric strain.

The hydrolysis of methyl benzoate and *p*-toluate in nearly 100 per cent sulphuric acid can also be interpreted in terms of an acyl carbonium ion. The hydrolyses are kinetically independent of the water concentration (below 1 M) and may be following the mechanism

$$PhCOOR \underset{fast}{\overset{H^+}{\rightleftharpoons}} PhCO\overset{+}{O}HR \xrightarrow{slow} PhCO^+ \xrightarrow[H_2O]{fast} PhCOOH.$$

The hydrolysis of β-lactones in dilute H_2SO_4 or $HClO_4$ involves acyl–oxygen fission, since when $H_2{}^{18}O$ enriched solutions are used, very little of the oxygen isotope appears in the β-hydroxyl group of the acidic product (equation 9.4).

$$\begin{array}{c} RCH-CH_2 \\ | \qquad | \\ O\!-\!CO \end{array} \xrightarrow{H^+/H_2{}^{18}O} \underset{\text{(mainly)}}{RCHOHCH_2CO^{18}OH} \qquad 9.4$$

The rate of the hydrolysis is also dependent upon h_0, and not upon the formal concentration of acid or of H^+; this dependence implies that it is the protonated lactone which is involved in the slow step of the hydrolysis

$$\underset{\underset{+}{H-O-CO}}{RCH-CH_2} \xrightarrow{\text{slow}} \underset{HO}{RCH-CH_2-\underset{+}{C}=O} \xrightarrow[H_2O]{\text{fast}} \text{product}$$

9.3.2 $A_{AC}2$ mechanism

Since esterification is a reversible process in acid conditions, a study of acid-catalysed hydrolyses of esters must provide us with some information about the mechanism of the reverse (esterification) reaction. The hydrolysis of γ-butyro-lactone, or of ethyl hydrogen succinate, proceeds with acyl–oxygen fission, for ^{18}O from the solvent is found in the carboxylic acid fragment, and not in the hydroxyl (alcohol) fragment. In contrast to the β-lactones, the rate of hydrolysis does not vary with h_0, but rather with $[H^+]$ in aqueous acid media.

The slow stage of the processes does not therefore involve the protonated substrate alone. If one molecule of water were also incorporated in the transition state, i.e.

$$\text{Rate} = k[\text{ester}][H^+][H_2O] = k'[\text{ester}-H^+][H_2O],$$

the observed rate would be proportional to the hydrogen ion concentration, $[H_3O^+]$, for the activity of H^+ in concentrated acids refers to the proton-donating abilities of *all* acids in the solution, and hence $[H_3O^+]$ would be simply linked to $[H_2O]$ and the acidity of the medium,

$$H_2O \underset{}{\overset{\text{`}H^{+}\text{'}}{\rightleftharpoons}} H_3O^+,$$

$$[H_3O^+] = K[H_2O]h_0 .$$

A mechanism which fits these observations would be

$$\underset{H}{\overset{R'}{RCOO^+}} \underset{\overset{-H_2O, \text{ fast}}{\longleftarrow}}{\overset{H_2O, \text{ slow}}{\longrightarrow}} \left[\underset{\underset{R'}{\overset{|}{O}}{\underset{H}{\nearrow}}}{\overset{\overset{O}{\underset{|}{\parallel}}}{R-C-OH_2}} \right]^+ \underset{\overset{R'OH, \text{ slow}}{\longleftarrow}}{\overset{-R'OH, \text{ fast}}{\longrightarrow}} RCOOH_2^+$$

in which the protonated ester or acid underwent slow nucleophilic attack by water or by an alcohol to provide a common intermediate whose subsequent decomposition was a relatively fast process. The intermediate, if it has some stability, may allow exchange and migration of hydrogen between the individual oxygen atoms attached to carbon (carboxylate carbon $-CO_2H$), when oxygen exchange with the solvent may appear to occur. This has, in fact, been found in the hydrolysis of $PhC^{18}OOEt$ in acid media. Similarly, acid-catalysed ester interchange (exchange reactions) would proceed through a similar intermediate which may lose either ROH or R'OH.

9.4   Alkyl–oxygen fission

9.4.1 $A_{AL}1$ *process*

Since the hydrolysis of inorganic esters of tertiary alcohols can proceed by a unimolecular mechanism (e.g. *t*-BuCl in aqueous acetone or aqueous alcohol), it would be expected that *t*-butyl acetate and its analogues could hydrolyse by a similar mechanism. In $H_2{}^{18}O$, labelled *t*-BuOH is formed, indicating the formation of *t*-Bu$^+$ or its kinetic equivalent. Using an ester of an optically active tertiary alcohol, one finds the derived alcohol to be racemized. Since a bimolecular displacement by solvent would cause inversion of configuration, we can be more certain of the unimolecular solvolysis mechanism, which would predict the formation of a planar carbonium ion and hence the racemization of an optically active alcohol fragment (equation 9.5)

$$\text{RCOOR}' \xrightleftharpoons{\text{slow}} (R')^+ \, \text{RCO}_2^- \xrightarrow{\text{fast, } H_2O} R'\text{OH} + \text{RCO}_2\text{H}. \qquad 9.5$$

An acid-catalysed hydrolysis could be similarly depicted, when protonation of the ester allowed heterolysis to occur even more readily.

Both acyl–oxygen and alkyl–oxygen fission are catalysed by acids. The two processes make opposite demands upon the ester, however, since the $A_{AC}1$ process (when RCO^+ and $R'OH$ are formed) would be favoured by electron-repelling groups in R and electron-attracting groups in R', while the $A_{AL}1$ process will be assisted by substituents stabilizing the R' carbonium ion (electron-repelling groups). A change-over in mechanism could therefore occur through a series of esters, and it is in this way that the rates of hydrolysis of the alkyl benzoates in sulphuric acid are explained (Figure 23).

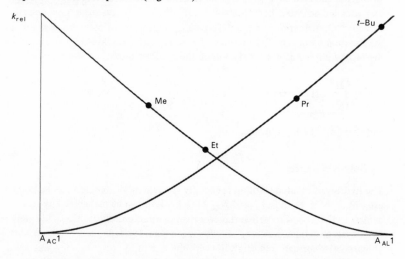

Figure 23 Relative rates of hydrolysis of alkyl benzoates in sulphuric acid

The order and explanation parallel that applied to the rates of solvolysis of the alkyl halides, when a transition from S_N2 to S_N1 behaviour occurs.

9.4.2 $B_{AL}1$ and $B_{AL}2$ mechanisms

In the absence of added acid, alkyl–oxygen fission of esters may still occur to give carbonium ionic intermediates,

$$R'COOR \rightleftharpoons R'COO^- + R^+,$$

and such hydrolyses will be characterized by rates of reaction which are independent of hydroxide ion concentration, and which yield racemic alcohols from optically active esters. Some substituted benzyl hydrogen phthalates (e.g. 9.2)

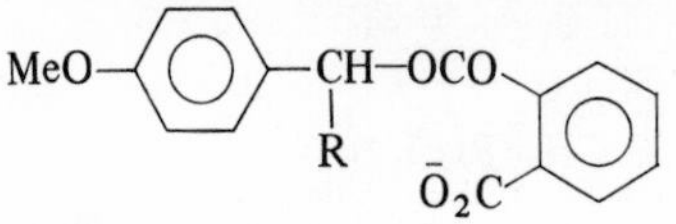

(9.2)

have shown such behaviour in dilute alkali. In concentrated alkali, however, a bimolecular process could obscure the observation of a unimolecular process. With only the most favourable conditions (R = p-MeO—$C_6H_4\overset{+}{C}HPh$) can the $B_{AL}1$ mechanism be found in concentrated (10 M) aqueous hydroxide; the other esters now show acyl–oxygen fission since the bimolecular process ($B_{AC}2$) has now overtaken the unimolecular one.

The $B_{AL}2$ process is rare for carboxylic esters. The hydrolysis of β-butyrolactone, although showing acyl–oxygen fission ($B_{AC}2$) in strongly basic media and also in strongly acidic media ($A_{AC}1$), appears to show a new mechanism in intermediate basicities and acidities. In ^{18}O-labelled water the β-hydroxyl group now does contain ^{18}O; furthermore, hydrolysis results in inversion of configuration about the β-carbon atom. The first result might merely indicate alkyl–oxygen fission, but the second implies attack by H_2O upon the β-carbon atom,

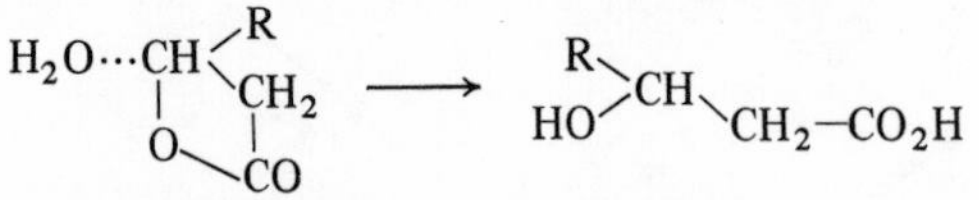

and therefore the $B_{AL}2$ mechanism.

9.5 Hydrolysis of amides

The hydrolysis of amides follows generally the two more common mechanisms of ester hydrolysis. Both $B_{AC}2$ and $A_{AC}2$ mechanisms can be recognized by tests similar to those used to define the processes in ester hydrolysis. The acid-catalysed hydrolysis of amides, however, shows a more complex dependence upon acidity (Figure 24). 'Normal' acid catalysis (with the rate of hydrolysis dependent upon h_0) continues until all the amide in solution is protonated, effectively. Higher concentrations of acid cannot further cause acceleration by forming the protonated

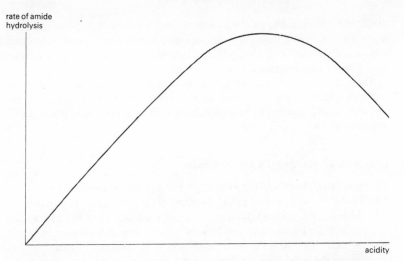

rate of amide hydrolysis

acidity

Figure 24 Dependence of acid-catalysed hydrolysis of amides upon acidity

amide, but rather cause a lowering in the concentration of water, the other reagent. As a result, the reaction is firstly accelerated by increasing acidity up to the point where $[\text{amide}-\text{H}^+] = [\text{amide}]_0$, and then slows down as $[\text{H}_2\text{O}]$ decreases with increasing acidity.

The appropriate mechanisms for acid- and base-catalysed hydrolysis will therefore be:

acid-catalysed

$$\text{amide} \; \underset{}{\overset{\text{H}^+}{\rightleftharpoons}} \; [\text{RCONH}_2\text{R}']^+$$

$$\text{R}-\overset{+}{\text{CO}}-\text{NH}_2\text{R}' + \text{H}_2\text{O} \; \underset{}{\overset{\text{slow}}{\rightleftharpoons}} \; \left[\begin{array}{c} \text{O} \\ | \\ \text{R}-\text{C}-\text{NH}_2\text{R}' \\ | \\ \text{OH}_2 \end{array} \right]^+$$

$$\downarrow \text{fast}$$

$$\text{RCO}_2\text{H} + \text{R}'\text{NH}_3^+ \; \underset{}{\overset{\text{fast}}{\rightleftharpoons}} \; [\text{RCOOH}_2]^+ + \text{R}'\text{NH}_2$$

base-catalysed

$$\text{R}-\overset{\delta+}{\text{C}}=\overset{\delta-}{\text{O}} + \text{OH}^- \; \underset{}{\overset{\text{slow}}{\rightleftharpoons}} \; \left[\begin{array}{c} \text{OH} \\ | \\ \text{R}-\text{C}-\text{O} \\ | \\ \text{NHR}' \end{array} \right]^- \; \overset{\text{fast}}{\longrightarrow} \; \begin{array}{c} \text{RCOOH} \\ {}^-\text{NHR}' \end{array}$$
$$\quad\; \text{NHR}'$$

$$\text{fast} \; \updownarrow$$

$$\text{RCO}_2^- + \text{R}'\text{NH}_2$$

Unlike the corresponding ester reactions, the amide hydrolysis is irreversible. In acid media, the amine fragment is removed from the sphere of action by protonation; in base, the carboxylic acid is deactivated by loss of a proton to form the anion. In each case the final product cannot re-enter into the reaction.

The evidence for the intermediate adduct in acid-catalysed hydrolysis is far less convincing than in ester hydrolysis. No oxygen exchange occurs between solvent and the reacting species, and hence the life of such an intermediate, if it is measurable, must be very small.

9.6 Hydrolysis of acyl chlorides and anhydrides

The rapid solvolysis of acyl chlorides suggests that both unimolecular and bimolecular mechanisms might exist, for some of the most sterically hindered acid chlorides (e.g. mesitoyl chloride, the corresponding ester of which is resistant to base-catalysed hydrolysis) react the most rapidly. In fact, both mechanisms may be observed in the substituted benzoyl chlorides. The unimolecular mechanism, shown by mesitoyl chloride in 95 per cent aqueous acetone, requires the presence of a polar solvent, and a substituent capable of stabilizing acyl carbonium ionic species, for its operation (equation 9.6)

$$\text{RCOCl} \underset{}{\overset{\text{slow}}{\rightleftharpoons}} \text{[RCO]}^+ \xrightarrow[\text{H}_2\text{O}]{\text{fast}} \text{RCOOH} + \text{H}^+. \qquad \textbf{9.6}$$

Under less advantageous conditions (e.g. p-NO_2 substituent, or solvolysis in ethanol–ether mixture) a bimolecular mechanism can be found. The rate of solvolysis is accelerated by added nucleophiles (e.g. OEt^-; cf. unimolecular process), and may be represented by the mechanism

Similarly, we would expect the possibility of a unimolecular solvolysis mechanism for acid anhydrides. As yet there has been insufficient work upon these systems to allow two mechanisms to be clearly defined.

Problems

9.1 The base-catalysed cleavage of β-diketones occurs readily in aqueous or in ethanolic solution. In aqueous solution, rate = $k[\text{OH}^-][\text{diketone}]$. In ethanolic media, the kinetics of the reaction depends upon which species is in excess. If the diketone is in excess, rate = $k[\text{OEt}^-]$; if ethoxide is in excess, the rate depends only upon the concentration of the diketone.

(a) Since these results were obtained using acetyl-acetone, show that the kinetic results in ethanol are consistent with the rate of reaction being dependent upon the concentration of a carbanion.

(b) $CH_3COCMe_2COCH_3$ is cleaved more rapidly than acetylacetone; is this consistent with a carbanion intermediate?

(c) Show that the kinetics are also consistent with a rate-determining attack by OEt^- (or OH^-) upon the β-diketone, and that formation of the carbanion is a retarding effect, since it removes diketone from the sphere of the reaction. (See Pearson *et al.*, 1951.)

9.2 The hydrolysis and alcoholysis of acid chlorides by a unimolecular mechanism has been suggested to involve the sequence

$$RCOCl \xrightarrow{\text{slow}} RCO^+ + Cl^-$$

$$RCO^+ + R'OH \xrightarrow{\text{fast}} RCOOR' + H^+$$

While this has obvious analogies to the S_N1 reaction mechanism of alkyl halides, is it the only possible mechanism?

Chapter 10
Molecular Rearrangements

In Chapter 4 the migration of substituents was found to occur during neighbouring-group participation. We will now consider a number of rearrangements, some of which have analogies with the simpler systems already discussed.

10.1 Wagner–Meerwein rearrangements

The rearrangement of camphene hydrochloride (10.1) to isobornyl chloride (10.2)

(10.1)   (10.2)

occurs readily in polar solvents ($MeNO_2$, liquid SO_2) and is catalysed by Lewis acids ($HgCl_2$, HCl, or $SnCl_4$) but not by halide ion. The rearrangement presumably therefore involves a carbonium ionic intermediate, since all the species which will accelerate the process assist in the ionization of camphene hydrochloride. Since ^{36}Cl is exchanged between labelled HCl and (10.1) at a rate much greater than that of the rearrangement, the slow stage of the process would not involve formation of the carbonium ion, but rather its rearrangement. The formation of isobornyl chloride from the

(A = Lewis acid) $(ACl)^-$

carbonium ionic species depicted is the slow stage. Such an intermediate is a *non-classical carbonium ion*, for it cannot be written with charge localized upon only one carbon atom, as a classical ion requires; it must be written non-classically if we are to suggest neighbouring-group participation by the migrating bond.

Compared with ordinary tertiary chlorides (e.g. *t*-butyl chloride; much less dramatic differences occur if 1-chloro-1-methylcyclopentane is taken as standard) camphene hydrochloride solvolyses much more rapidly (by factors of 300 to 8000 times), which suggests participation, presumably by the electrons of the migrating bond; similarly isobornyl chloride reacts more rapidly than 'normal' secondary chlorides. Participation could occur in both cases *if* the finally produced carbonium ion were a non-classical structure (10.3).

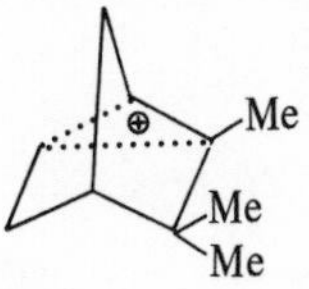

(10.3)

Similarly, a 1,3-hydride shift may occur, for the acetolysis of the norbornyl *p*-bromobenzenesulphonate (10.4) provides two products (10.5) and (10.6) in which a Wagner–Meerwein rearrangement and hydrogen shift have occurred.

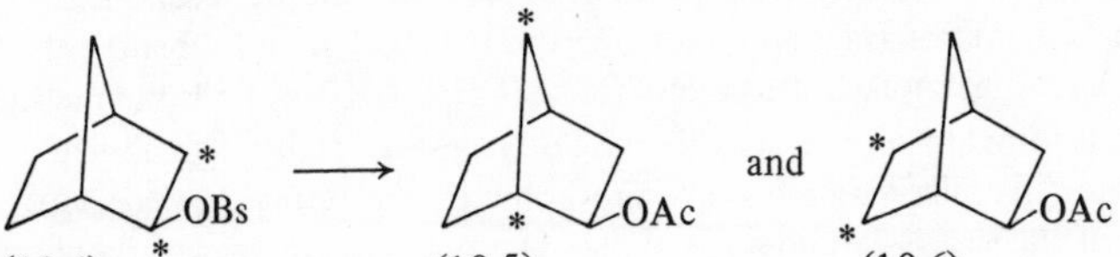

(10.4)   (10.5)   and   (10.6)

* = isotopic carbon.

10.2 The pinacol rearrangement and related reactions

The acid-catalysed dehydration of a 1,2-glycol may yield a rearranged product by the so-called 'pinacol rearrangement'. Pinacol itself (2,3-dimethyl-2,3-butanediol) may form either pinacolone (methyl *t*-butyl ketone) or 2,3-dimethyl-1,3-butadiene,

$$Me_2 C(OH)C(OH)Me_2 \xrightarrow{H^+} Me_3 CCOMe +$$
$$CH_2=C(Me)C(Me)=CH_2 ,$$

and in general the rearrangement occurs preferentially. Since acid conditions are necessary, a carbonium ionic intermediate seems probable; this intermediate could undergo alkyl group migration to give the conjugate acid of the ketone or aldehyde product,

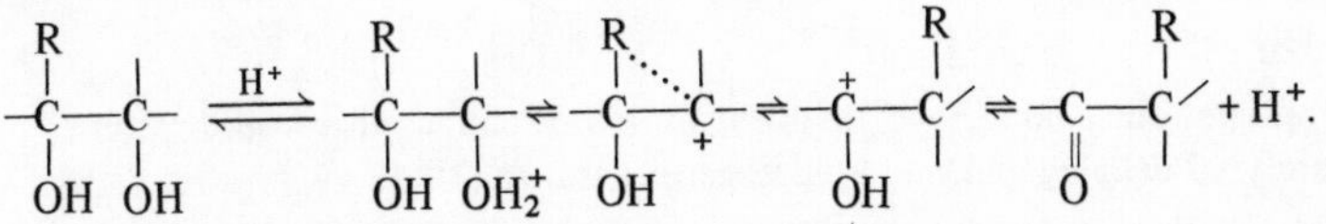

From such a mechanism we might expect conformational requirements similar to those found in neighbouring-group participation (section 4.3.1). *trans*-1,2-Dimethyl-1,2-cyclohexanediol undergoes rearrangement to form 1-acetyl-1-methylcyclo*pentane* (equation **10.1**), although the *cis* isomer, in which the methyl and the hydroxyl groups are in *trans*-diaxial configurations to each other, provides 2,2-dimethylcyclohexanone (equation **10.2**).

$$\text{(pinacol, two HO and two Me groups on ring)} \xrightarrow{\text{H}^+} \text{Me}-\text{C}(=\text{O})\text{Me (ketone)} \qquad \textbf{10.1}$$

$$\text{(HO, Me substituted ring)} \xrightarrow{\text{H}^+} \text{(2,2-dimethylcyclohexanone)} \qquad \textbf{10.2}$$

In deciding which substituent migrates in a given pinacol there are two conflicting factors to be considered. The first is the relative basicities of the two —OH groups, for this will decide the position of protonation; the second is the relative ease of migration (migratory aptitude) of the alkyl or aryl substituents. In 1,1-dimethyl-2,2-diphenyl-1,2-ethanediol ($Ph_2C(OH)C(OH)Me_2$) it is the methyl group which is found to migrate, not because of a superior ability to stabilize positive charge (the migrating group acquires some charge during the reaction), but because of the superior ability of the phenyl groups to stabilize a carbonium ion, hence ensuring the position of protonation of the pinacol.

A series of migratory aptitudes have been measured for a number of tetra-aryl-pinacols ($Ar(Ar')C(OH)C(OH)Ar(Ar')$) for which the protonation of either OH group occurs equally readily. The order of migration of substituted phenyl systems, *p*-OMe > *p*-Me > *p*-Ph > H > *p*-Cl > *o*-OMe, parallels the ability of the system to stabilize positive charge. The very low aptitude of the *o*-OMeC$_6$H$_4$— system presumably reflects the considerable crowding in the intermediate ions, when steric factors outweigh electronic effects.

10.2.1 *Deamination of β-amino-alcohols*

As aliphatic diazonium ions are generally unstable, treatment of a β-amino-alcohol with nitrous acid will provide a carbonium ionic species similar in structure and reaction to the first intermediate in the pinacol rearrangement (equation **10.3**).

$$-\underset{\underset{NH_2}{|}}{C}-\underset{\underset{OH}{|}}{C}- \xrightarrow{\text{HONO}} -\underset{\underset{N_2^+}{|}}{C}-\underset{\underset{OH}{|}}{C}- \xrightarrow{-N_2} -\underset{+}{\overset{|}{C}}-\underset{\underset{OH}{|}}{C}- \qquad \textbf{10.3}$$

Similar migrations could therefore occur in this system, as also would result in the Ag$^+$-catalysed dehalogenation of β-halogeno-alcohols (**10.4**).

$$-\underset{\underset{X}{|}}{C}-\underset{\underset{OH}{|}}{C}- \xrightarrow{\text{Ag}^+} -\underset{\underset{AgX}{\cdot\cdot+}}{C}-\underset{\underset{OH}{|}}{C}- \qquad \textbf{10.4}$$

The amino-alcohol (10.7), in which one of the phenyl groups is isotopically labelled, shows some surprising results. Labelled phenyl groups migrate mainly with *inversion* of configuration about C_α, whereas unlabelled phenyl migrates mainly with retention of configuration about this atom. The implication of this is that not only the conformation of the parent amino-alcohol and the initially formed carbonium ion are confined to one essential structure, but also that migration occurs very readily after carbonium ion formation has occurred. If the carbonium ion (10.8) were free to rotate before migration occurs, we would expect migration of labelled phenyl with retention of configuration, and migration of unlabelled phenyl with inversion of configuration about C_α. Neither of these are observed, and the migration of unlabelled Ph with retention of configuration (corresponding to $60°$ rotation about $C_\alpha - C_\beta$) accounts for 12 per cent of the reaction product.

The implication is that migration occurs either simultaneously with the formation of the carbonium ion, or at least very soon afterwards.

In the corresponding reaction of the simple amine $PhCH(Me)CH(Me)NH_2$ with nitrous acid, there is evidence of hydride shift ($PhCX(Me)CH_2Me$ product), and methyl shift ($PhCHXCHMe_2$ product) as well as phenyl group migration. As the extents of these migrations are roughly similar, and quite different to those expected from the relative abilities of the groups to stabilize positive charge, it seems that the loss of nitrogen from the diazonium ion occurs without any assistance from groups attached to the β-carbon atom, and that which group migrates depends solely upon the conformation of the classical carbonium ion which is formed. A similar situation seems to occur in the rearrangements of the β-amino-alcohols.

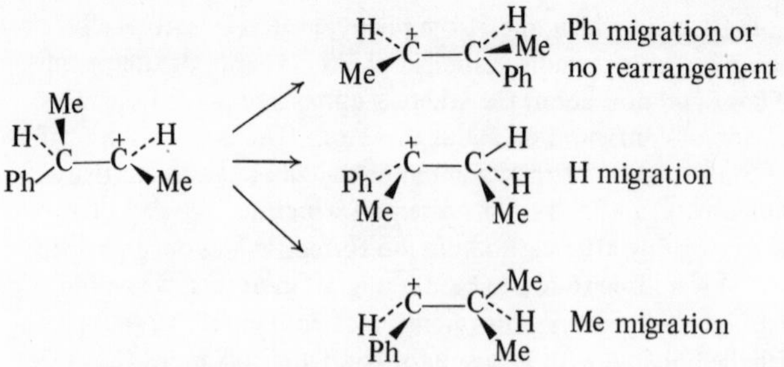

Ph migration or no rearrangement

H migration

Me migration

10.3 Rearrangements involving nitrogen atoms

The previous examples have involved migrations to electron-deficient carbon atoms; the present series of reactions involve migration to nitrogen.

10.3.1 *Beckmann rearrangement*

The formation of an amide from a ketoxime occurs on treatment with acidic species (H_2SO_4, P_2O_5, SO_3) or with reagents which will form more labile esters of the oxime (RSO_2Cl, PCl_5, $SOCl_2$). The acid-catalysed process involves the protonated oxime, since the rates of rearrangement parallel h_0 in aqueous sulphuric acid solutions; the rearrangements of the oxime esters presumably involve heterolysis of the N—X bond, for the reaction is kinetically first-order in ester and is accelerated by increased solvent polarity (equation **10.5**).

$$\text{\textbackslash}C=NOH \overset{H^+}{\rightleftharpoons} [\text{\textbackslash}C=NOH_2^+] \overset{slow}{\longrightarrow} \text{product (} \text{\textbackslash}CONH-)$$

$$\text{\textbackslash}C=NX \rightarrow -C(X)=N- \overset{H_2O}{\longrightarrow} -CONH- + HX$$

10.5

The migrating group has been shown to approach the nitrogen atom from the opposite side to that from which the group X is expelled (a parallel with the pinacol rearrangement) and that the migrating carbon atom retains its configuration. This last point was shown by the retention of optical activity when the amide (10.10) was formed from the optically active oxime (10.9).

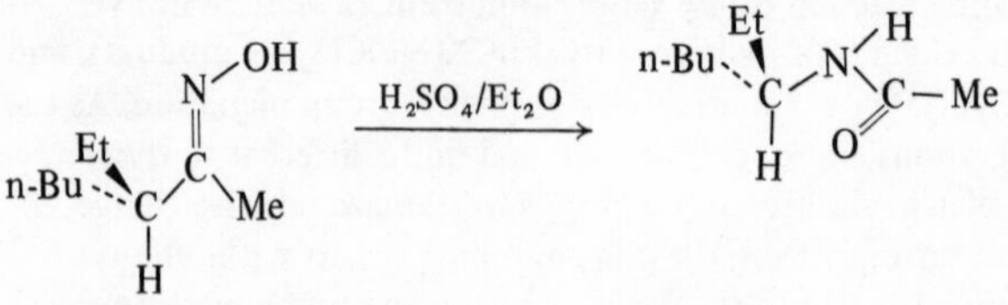

(10.9)   (10.10)

While it is uncertain whether migration or loss of the leaving group (X or H_2O) occurs first, it is now definite that the two processes are almost simultaneous, since the leaving group is evidently shielding one side of the carbon–nitrogen multiple bond.

$$(10.9) \longrightarrow \qquad \longrightarrow \qquad \longrightarrow (10.10)$$

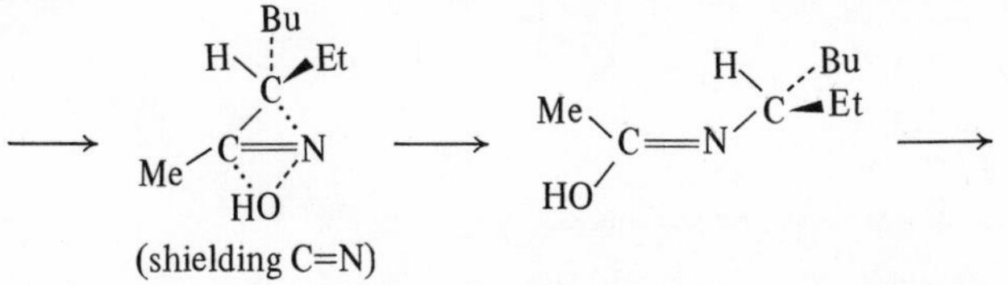

(shielding C=N)

3.2 Hofmann, Curtius, Schmidt, and allied reactions

The formation of a primary amine from a carboxylic acid may employ a number of acyl derivatives, all of which finally may form an alkyl isocyanate whose hydrolysis provides the amine.

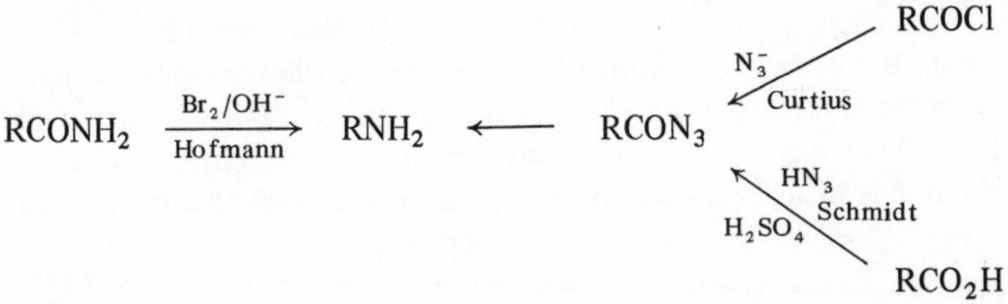

$$RCONH_2 \xrightarrow[\text{Hofmann}]{Br_2/OH^-} RNH_2 \longleftarrow RCON_3$$

with Curtius (from RCOCl, via N_3^-) and Schmidt (from RCO_2H, via HN_3, H_2SO_4) routes shown to $RCON_3$.

In the *Hofmann rearrangement*, an N-halogenoamide (readily prepared by direct halogenation of amides) gives an isocyanate on treatment with base. The first stage is presumably the formation of the anion RCONBr, for N-alkyl-halogenoamides do not undergo the reaction. Migration of the alkyl or aryl group, R, probably occurs as the halide ion is expelled; it has been suggested that if loss of X^- occurred before migration, the resulting structure (RCON=) would react readily with the solvent to give hydroxamic acids, which are not found.

Similarly, we would expect the migrating carbon atom to retain its configuration, as in the Beckmann rearrangement. This is shown by the ready formation of bridge-head amines such as (10.11), when the rigidity of the system does not allow conformational rearrangements.

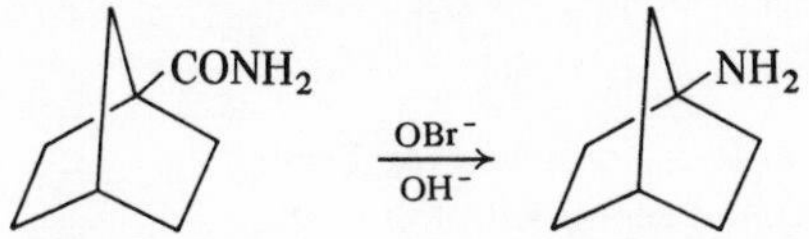

$$\xrightarrow[\text{OH}^-]{OBr^-}$$

(10.11)

Similarly, camphoric acid amide (10.12) provides an amino-acid which readily forms an internal amide (lactam). The amino- and carboxylic acid groups must therefore lie on the same side of the cyclopentane ring, and so no inversion can have occurred during the Hofmann reaction.

Me $\diagdown$ Me ⟶ Me $\diagdown$ Me ⟶ Me $\diagdown$ Me

Me, bicyclic structure with —CONH$_2$ and —COOH groups

$$\text{(bicyclic)}-\overset{\text{Me}}{}\overset{\begin{array}{c}\text{CONH}_2\\\text{COOH}\end{array}}{} \xrightarrow[\text{OH}^-]{\text{OBr}^-} \text{(bicyclic)}\overset{\begin{array}{c}\text{Me}\\\text{NH}_2\\\text{COOH}\end{array}}{} \xrightarrow{\text{readily}} \text{(bicyclic)}\overset{\begin{array}{c}\text{Me}\\\text{NH}\\\text{C}{\diagdown}\text{O}\end{array}}{}$$

(10.12)

The mechanism proposed for this reaction,

$$\text{RCONHX} \underset{}{\overset{\text{OH}^-}{\rightleftharpoons}} \text{RCON}^-\text{X} \xrightarrow{\text{slow}} \underset{\overset{\|}{O}\ \overset{}{X}}{\overset{R}{C}-\overset{..}{N}} \xrightarrow{-X^-} \text{RN}{=}\text{C}{=}\text{O} \to \text{RNH}_2$$

is consistent with its acceleration by electron-donating groups in the aromatic ring (R = Ph); the intermediate presumably has some measure of shielding at one side of the C—N bond, so that migration can occur with retention of configuration.

The decomposition of acyl azides ($RCON_3$) to form isocyanates after rearrangement is the common feature of the *Curtius* and the *Schmidt rearrangements*. Again the group migrates with retention of configuration and again the loss of nitrogen and the migration appear to be nearly coincident. Surprisingly, the Curtius reaction is an acid-catalysed process. Since the rate-determining stage of the reaction involves breaking the bond between N_α and N_β,

$$\underset{\overset{\|}{O}}{R-C}-\overset{+}{N}{=}\overset{}{N}{=}\overset{-}{N} \qquad \underset{\overset{\|}{O}}{R-C}\overset{\frown}{\text{...}}\overset{+}{N}\text{...}N{\overset{..}{=}}N$$
$$\qquad\quad \alpha\ \ \beta\ \ \gamma$$

this can be explained by suggesting that protonation of carbonyl oxygen, by allowing the π-electrons of N_α to be drawn away from those of N_β, facilitates the heterolysis of the N—N bond by reducing its multiple character.

The *Schmidt* reaction, applied to carboxylic acids, presumably involves a similar intermediate,

$$\text{RCO}_2\text{H} \underset{}{\overset{H^+}{\rightleftharpoons}} \text{R}\overset{+}{\text{C}}(\text{OH})_2 \underset{}{\overset{\text{HN}_3}{\rightleftharpoons}} [\text{RC(OH)}{=}\text{NN}{\equiv}\text{N}]^+$$
$$\downarrow \text{slow}$$
$$\text{RNH}_2 \xleftarrow{\text{H}_2\text{O}} \text{R}-\text{N}{=}\overset{+}{\text{C}}-\text{OH} + \text{N}_2$$

whose decomposition, with migration of R, is the slow stage forming the conjugate acid of the isocyanate RNCO. In keeping with this mechanism, electron-donating groups in R facilitate the process, both by stabilizing intermediate ions and by their more favourable migratory aptitudes.

The migration, as with the Hofmann and the Beckmann rearrangements, occurs with retention of configuration and presumably involves similarly rigid conformational species as intermediates.

On treatment of other carbonyl derivatives (aldehydes, ketones) with HN_3 in acidic media, similar rearrangements occur to give amides or allied species.

4 Aromatic rearrangements

The so-called 'aromatic' rearrangements generally involve the migration of groups initially attached outside of the phenyl ring system to direct attachment to the ring. The benzidine rearrangement, as well as a number of rearrangements involving N- or O-substituted aromatic systems, fall under this heading.

4.1 Benzidine rearrangement

The reaction of hydrazobenzenes with dilute acids causes isomerization. If the *para* position is vacant, 4,4'-diaminobiphenyls are formed; when one such position is successfully blocked,* derivatives of 2,4'-diaminobiphenyl ('diphenylenes'), 2-amino- (*ortho*-semidine) or 4-amino-diphenylamine (*para*-semidine) may result.

The reaction is an intramolecular process, for the rearrangement of two hydrazo-benzenes (e.g. di-*o*-methoxy- and di-*o*-ethoxyhydrazobenzenes) does not give a mixed product, showing that the intermediates in the reaction do not become free (equation **10.6**).

$$\left. \begin{array}{c} \text{H–Ar–NHNH–Ar–H} \\ \text{and} \\ \text{H–Ar}'\text{–NHNH–Ar}'\text{–H} \end{array} \right\} \xrightarrow{\text{H}^+} \left\{ \begin{array}{l} \text{H}_2\text{N–ArAr–NH}_2 \text{ and} \\ \text{H}_2\text{N–Ar}'\text{Ar}'\text{–NH}_2 \\ \text{but no } \text{H}_2\text{N–Ar}'\text{Ar–NH}_2 \end{array} \right. \qquad \textbf{10.6}$$

Kinetically, the rate of rearrangement of di-*p*-deuterohydrazobenzene is almost that of the unlabelled isomer, and the overall process is third order,

Rate = k[hydrazobenzene] $[\text{H}^+]^2$.

The displacement of the aromatic proton (in the *para* position) is therefore not kinetically significant, and the di-cation of hydrazobenzene, or its kinetic equivalent, seems to be involved in the slow stage of the process.

* A *p*-substituent may be displaced during the rearrangement; for instance, –D, –SO₃H, –CO₂H, and –I may all be expelled during the formation of a 4,4'-diaminobiphenyl.

A sequence which fits these results is

$$\text{(C}_6\text{H}_5\text{)}-\text{NHNH}-\text{(C}_6\text{H}_5\text{)} \underset{\text{fast}}{\overset{2H^+}{\rightleftharpoons}} \left[\ldots \right]^{2+} \xrightarrow{\text{slow}} \ldots$$

$$\xrightarrow[\text{fast}]{-2H^+} \quad H_2N-\text{(C}_6\text{H}_4\text{)}-\text{(C}_6\text{H}_4\text{)}-NH_2$$

in which the two aromatic systems lie parallel to each other and the fragments, after fission of the N–N bond, rotate through relative angles of $60°$ (giving *o*-semidine), through $180°$ (giving *p*-semidine), or $120°$ (giving 2,4′-diaminobiphenyl). The intermediates proposed in this mechanism are not wholly satisfactory. The two aromatic rings will suffer considerable internal strain (apart from that imposed upon the C–N and N–N bonds) in forming the 'sandwich' intermediate, particularly as a di-cation; di-*p*-xylylene, in which two benzene rings are held together by two $-CH_2-CH_2-$ groups, is definitely not two planar phenyl systems held by aliphatic chains, but shows considerable warping of the aromatic rings. The intermediate in the Benzidine rearrangement is a similar structure in which conjugation can apparently take place through the (non-planar) phenyl rings from the exocyclic nitrogen atom. However, this is not, in the author's opinion, reason for discarding such an intermediate, for substituents in one ring of a *para*-cyclophane (of which di-*p*-xylylene is a member) may influence orientation of attack at the unsubstituted ring (indicating an electronic effect outside of the one phenyl system) and may even result in carbon–carbon bonding between the rings. The nitration of *meta* [2,2] cyclophane gives 2-nitro-4,5,9,10-tetrahydropyrene with such bond formation (equation **10.7**) and so the possibility of forming an

$$\text{[meta[2,2]cyclophane]} \xrightarrow{HNO_3} O_2N-\text{[2-nitro-4,5,9,10-tetrahydropyrene]} \qquad \textbf{10.7}$$

interannular bond, or of electronic effects from one phenyl ring to another, is not precluded if the aromatic systems are not strictly planar. The formation of such an intermediate as (10.13) is the most tenuous part of the argument,

(10.13)

because of the unfavourable steric and electronic effects. Various attempts have been made to explain the stability of this intermediate in terms of hyper-conjugation or some excess resonance stabilization. The problem might not be real, for the kinetics of the reaction are consistent with proton attack upon a mono-protonated species (e.g. 10.14) in the rate-determining stage, or with any

$$\xrightarrow{H^+}$$ intermediate.

(10.14)

other kinetically equivalent process. Once the N–N bond is broken, however, further bond formation must occur very quickly, for the two similarly charged (mutually repulsive) species never become free from each other, since 'mixed' benzidines are not formed.

4.2 Orton rearrangement and related reactions

The treatment of N-chloroacetanilides with acid (particularly HCl) causes apparent rearrangement with the production of o- and p-chloroacetanilides. Rearrangements using HCl as the mineral acid are third-order processes,

$$\text{Rate} = k_3 [\text{HCl}]^2 [N\text{-chloroacetanilide}],$$

but in contrast with the Benzidine rearrangement, a more nucleophilic substrate in the solution will be chlorinated preferentially. Anisole or acetanilide both yield mono-chloro-derivatives if they are present in solutions of $N,2,4$-trichloro-acetanilide and HCl. The kinetics are consistent with a rate-determining attack by Cl^- upon the protonated amide,

$$\begin{aligned}
\text{Rate} &= k [\text{protonated amide}] [Cl^-] \\
&= k' [H^+] [\text{amide}] [Cl^-] \\
&= k_3 [\text{HCl}]^2 [\text{amide}],
\end{aligned}$$

since $\quad HCl \rightarrow H^+ + Cl^-\quad$ and hence $\quad [H^+] = [Cl^-] = K[\text{HCl}],$

which would imply that the halogenating agent is in fact free chlorine. This would explain the halogenation of other reactive species in the rearrangement solution, and the removal of molecular chlorine from such solutions by a current of air. Similarly, HBr or HI, when used as proton sources, cause bromo- or iodo-acetanilides to be formed through the intermediacy of BrCl or ICl. Probably the rearrangement involves the sequence

for the isomer ratio of the chloroacetanilide product mixture is also that found from halogenation using molecular chlorine directly.

The rearrangement of N-nitroamines (e.g. $PhNHNO_2$) in acid media might seem analogous to that of the N-chloroamides. In fact, no nitrating species (such as NO_2^+ or HNO_3) is liberated in the course of the rearrangement, for there is no exchange with labelled nitrate ion in solution, nor are more readily attacked species (e.g. $PhNMe_2$) substituted in such solutions. Finally, while rearrangement gives almost exclusively o-nitroaniline from $PhNHNO_2$ in 85 per cent H_2SO_4, the direct nitration of aniline under these conditions gives a mixture of isomers of which o-nitroaniline only constitutes 6 per cent.

While this reaction is obviously an intramolecular process, and acid-catalysed, its mechanism is not obvious.

In contrast, and for similar reasons to those given for the Orton rearrangement, the reactions

$$ArN=NNHAr \xrightarrow{H^+} p\text{-aminoazobenzene}$$

and $ArN-NO \xrightarrow{H^+} p\text{-nitroso-}N\text{-alkylaniline}$
$\quad\quad |$
$\quad\quad R$

are both intermolecular processes, with the free diazonium ion and free nitrous acid as the respective intermediates.

Problems

10.1 Both isomers of 7,8-diphenylacenaphthene-7,8-diol (10.15) react with acid to

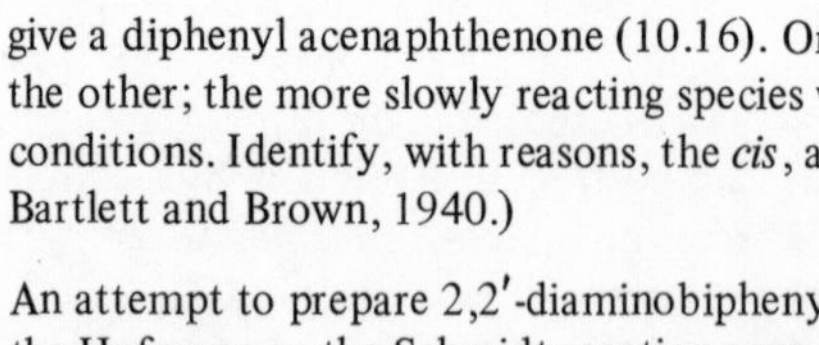

(10.15)   (10.16)

give a diphenyl acenaphthenone (10.16). One diol reacts at six times the rate of the other; the more slowly reacting species will isomerize under the reaction conditions. Identify, with reasons, the *cis*, and the *trans* forms of the diol. (See Bartlett and Brown, 1940.)

10.2 An attempt to prepare 2,2'-diaminobiphenyl from diphenic acid through either the Hofmann or the Schmidt reaction gave poor yields of base, although little acidic material could be recovered. The main product in each case was non-basic and non-acidic. Suggest a structure for it, and explain its formation.

0.3   The benzilic acid rearrangement involves a shift of an aryl group in the α-diketone $ArCOCOAr$ on treatment with base (OH^-) to form the anion of a benzilic acid, $Ar_2C(OH)CO_2H$. Suggest a way in which this rearrangement could reasonably occur, and try to account for the formation of $ArCOCPh_2OH$ from the Grignard reaction between $ArMgBr$ and benzil ($Ar = o$-tolyl). (See Westheimer, 1936, and Roberts and Urey, 1938.)

0.4   In acid media $(Ac_2O–H_2SO_4)$, dienones rearrange to phenol esters. Typically, (10.17) gives 3,4-dimethyl-1-naphthol as its acetate. What is a probable rearrangement mechanism? Does your mechanism predict that (10.18) will give 3-phenyl-4-methyl-1-naphthyl acetate and not the isomer?

   Me   Me   Me   Ph
   (10.17)   (10.18)

(See Arnold and Buckley, 1949.)

Chapter 11
Free-Radical Processes

11.1 Introduction

The processes which have been considered so far have all involved charged, or dipolar, intermediates and have been characterized by heterolytic bond-breaking processes. Reactions in which free radicals (atoms with unpaired electrons) are involved are common (e.g. auto-oxidation of fats and oils) but usually more complex than the heterolytic counterparts. For instance, the addition of chlorine to an olefin heterolytically provides almost entirely the corresponding dichloro-adduct,

$$RCH=CH_2 + Cl_2 \xrightarrow{\text{HOAc}} RCHClCH_2Cl,$$

whereas the attack of $Cl\cdot$ upon the same olefin will result in addition, and hydrogen abstraction (mainly, but not entirely, from the allylic carbon atom adjacent to the double bond). Side reactions, such as dimerization of intermediate organic radicals, will also complicate the process.

11.2 Evidence for free radicals

The existence of stable radicals such as $Ph_3C\cdot$ has been known for seventy years. A host of triarylmethyl radicals have been prepared by dehalogenation of the halide Ar_3CX with metal (Zn, Hg, Ag), and their reactions with iodine, nitric oxide, and with other radicals have been well documented. Analogous species have been found involving nitrogen (e.g. pentaphenyl-pyrrole aminium ion radical (11.1), and the triphenylamine analogue (11.2), or oxygen (e.g. semiquinones (11.3)).

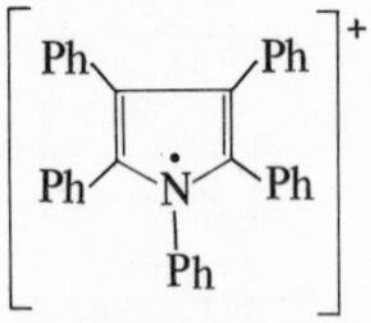

(11.1)   (11.2)   (11.3)

In all cases the stability of the radical is reflected by the large number of resonance forms which it is possible to write for the structure; the free electron may be located on a large number of carbon atoms in a species such as $Ph_3C\cdot$, and the more stable tris(*p*-biphenylyl)methyl radical, and such delocalization undoubtedly contributes to the stability of the system.

Short-lived radicals, such as Me· or Ph·, may exist as intermediates. We have already dealt with their means of detection and analysis (section 2.11.4), of which electron-spin resonance (e.s.r.) spectroscopy is the most valuable; they may be formed, as in the classic Paneth experiments, by the pyrolysis of organometallic species (e.g. $PbMe_4 \rightarrow$ Me·) and detected by the removal of mirrors of metal further down the tube.

$$PbMe_4 \xrightarrow{\text{heat}} Pb + 4Me \cdot \begin{cases} \longrightarrow & Me—Me \text{ (ethane)} \\ \longrightarrow & MeH \text{ (methane)} \\ \longrightarrow & CH_2{=}CH_2 \\ \xrightarrow{Pb, Sn, etc.} & PbMe_4, SnMe_4, etc. \end{cases}$$

Pyrolysis is probably the most important single means of producing free radicals. Organometallic species such as $PbEt_4$, and all those having covalent metal-carbon bonds are susceptible to thermal homolysis of the bond to give an organic radical. A second group of compounds which are used extensively to provide organic radicals upon heating are azo-compounds. The driving force for these reactions is generally the high stability of molecular nitrogen; CH_2N_2, $CH_3N{=}NCH_3$, and the host of aromatic species such as the covalent diazonium species $ArN{=}NX$ and $ArN{=}NCPh_3$ all provide radicals thermally by this means. Organic peroxides also are widely used as sources of radicals. The initially formed acyloxy radical ($RCOO \cdot$) loses carbon dioxide to form the radical R· (R may be alkyl or aryl; benzoyl peroxide is extensively used as a source of Ph·) which then undergoes conventional radical reactions:

$$(RCOO)_2 \xrightarrow{\text{heat}} 2RCOO \cdot \xrightarrow{-CO_2} 2R \cdot$$

All of these processes involve homolysis of a weak bond by thermal energy; it is not surprising that the same effect may be produced by irradiation with u.v. light, which has the advantage of restricting thermal decomposition of the products.

Other methods of formation, apart from pyrolysis, include photochemical decomposition and electrolysis. The Kolbe synthesis of hydrocarbons by the electrolysis of carboxylic acid salts relies on the formation of alkyl radicals by anodic oxidation. These species may then dimerize ($2R \cdot \rightarrow R—R$), disproportionate

$$RCOO^- \xrightarrow{\text{anode}} RCOO \cdot \xrightarrow{-CO_2} R \cdot$$

(giving olefins and H·), or react with other species in solution ($R \cdot + H \cdot \rightarrow RH$) to give the observed reaction products. Alkyl radicals are similarly formed by the electrolysis of Grignard reagents,

$$RMgBr \xrightarrow{\text{anode}} R \cdot + MgBr^+,$$

and a number of reductive dimerizations, such as pinacol formation from ketones, may be carried out under electrolysis conditions, when the pinacol precursor is probably a radical such as $R_2\dot{C}—O^-$.

Photochemical decomposition of species containing weak interatomic bonds (e.g. peroxides, organic iodides) is also an important source of radicals; since the initial decomposition is brought about by light and not by heat, the reaction products are less likely to suffer subsequent decomposition. Finally, radicals may be generated in solution as a result of one-electron oxidation processes. Fenton's reagent (Fe^{2+} and H_2O_2) supplies hydroxyl radicals,

$$Fe^{2+} + H_2O_2 \rightarrow Fe(OH)^{2+} + HO\cdot,$$

and similar electron transfers involving ferrous ion or titanium ions (Ti^{3+}) produce radicals from peroxydisulphate ion,

$$M^{n+} + S_2O_8^{2-} \rightarrow M(SO_4)^{n+} + [\cdot OSO_3]^{2-},$$

and from other intermediates. In all cases the fission of an O—O bond is necessary in forming the radical.

11.3   Reactions of free radicals

The probable reactions of free radicals have been adumbrated in previous sections. *Disproportionation* produces a saturated and an unsaturated species,

$$2\ RCH_2CH_2\cdot \rightarrow RCH_2CH_3 + RCH{=}CH_2$$

and represents a way in which radicals may be destroyed; *dimerization* is also a means of removing radicals from a system. Since radicals, when formed, may attack a non-radical and produce a new species which is also a radical and also capable of similar attack, it is not difficult to envisage a *chain* process in which one radical initiates a sequence whereby many molecules undergo attack,

e.g. $R\cdot + {-}CH{=}CH_2 \rightarrow -\dot{C}HCH_2R \rightarrow \cdots \rightarrow -\dot{C}HCH_2(CHCHR)_nH.$

In contrast with heterolytic processes, therefore, a free-radical reaction may be initiated by light, or by the presence of adventitious catalysts, and it may be inhibited or stopped by species which can bind up the radical intermediates in a non-reactive form. Iodine, or aromatic disulphides, are examples of inhibitors. Oxygen, a diradical, may function as a chain initiator (as in the autoxidation of oils) but may also behave as an inhibitor in a reaction which requires the periodic regeneration of a relatively reactive radical, the oxygen adduct of which ($R\cdot + O_2 \rightarrow ROO\cdot$) may be insufficiently reactive to serve as a substitute.

The relative insensitivity of homolytic reactions to solvent effects or to changes in the acidity of the medium (unless, of course, radical-ions are involved) has already been mentioned (section 2.11.4).

A homolytic reaction, therefore, can be divided into three stages. In the first, radicals are formed in an *initiation stage.* The radicals may then be involved in a *propagation sequence* in which the main reaction sequence (e.g. addition of HBr to an olefin) is carried out and the initiating free radical is regenerated, so that the resultant series of reactions are repeated many times (chain process). Finally, the chain is *terminated* by removal of the radical, usually by dimerization or disproportionation with another radical. In some sense, the radical may be regarded as a catalyst, introduced to and removed from the system in the initiation and termination steps.

In calculating the kinetic form of such processes, the 'steady-state' approximation is usually made. The concentration of all radical intermediates are thought to build up, very quickly, to a small value which remains invariant throughout the course of the reaction. If this is true, then the rate of formation of these intermediates must become zero very soon after the reaction has begun, and a number of equations may be set up which will allow the rate of the reaction (i.e. the rate of formation of products or the rate of removal of starting materials) to be evaluated. For instance, in the process

$$A \rightarrow R\cdot \qquad \text{(initiation)},$$

$$\left.\begin{array}{l} R\cdot + Cl_2 \rightarrow RCl + Cl\cdot \\ Cl\cdot + RH \rightarrow HCl + R\cdot \end{array}\right\} \quad \text{(propagation)},$$

$$\left.\begin{array}{l} 2R\cdot \rightarrow RR \\ R\cdot + Cl\cdot \rightarrow RCl \end{array}\right\} \quad \text{(termination)},$$

$$\frac{d[R\cdot]}{dt} = k_1[A] - k_2[R\cdot][Cl_2] + k_3[Cl\cdot][RH] - k_4[R\cdot]^2 - k_5[R\cdot][Cl\cdot] = 0.$$

A similar equation can be set up for the changes in $[Cl\cdot]$. The rate of removal of the reagent, A, or the rate of formation of either product (HCl or RCl) could then be evaluated.

.4 Homolytic addition reactions

The 'anti-Markovnikov' direction of addition of HBr to substituted olefins,

$$RCH=CH_2 + HBr \rightarrow RCH_2CH_2Br,$$

has been shown to be due to a homolytic process. Other additions, catalysed by free-radical sources, include halogen addition to unsaturated systems and also the polymerization of olefins such as ethylene and propylene, a reaction of considerable industrial importance.

.4.1 *Addition of hydrogen halides*

For some while the 'peroxide effect' has been the explanation of the unexpected direction of addition of HBr to allyl halides and other unsymmetrically substituted olefins. Even small quantities of organic peroxides (derived from oxygen and the olefin) will initiate a process in which hydrogen abstraction from HBr leaves $Br\cdot$ ($RO{-}OR \rightarrow 2RO\cdot$; $2RO\cdot + 2HBr \rightarrow 2ROH + Br\cdot$ which will then form a secondary radical $RCHCH_2Br$ (more stable than $RCHBrCH_2\cdot$; cf. section 8.2.1 – Markovnikov rule) also capable of abstracting hydrogen from HBr and providing the chain initiator $Br\cdot$ again.

Interestingly, these free-radical additions are stereo-specifically *trans*-processes, although a bromonium ion analogue (11.4) is unlikely on the grounds of its unusual number of valence electrons.

$$\overset{\times\times}{\underset{\times}{\times}}C \;\overset{\times}{\cdot}\; \ddot{C}{:}$$

$$\overset{\circ}{\underset{\circ\;\circ\;\circ}{\circ}}\overset{\bullet\bullet}{Br}{\overset{\circ}{\circ}}$$

(11.4) ⟋ 9 electrons

An explanation may rely on the axial attack on the double bond by Br·, giving a strained radical in which pyramidal rather than planar configuration around the tervalent carbon atom is found. Hydrogen abstraction from HBr would then be stereospecific, giving a *trans* adduct and regenerating Br·. Other explanations have been given, for olefin attack from the axial direction seems subject to steric hindrance.

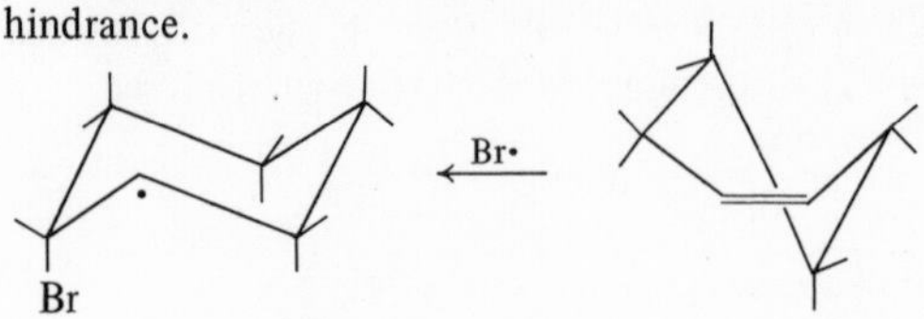

For non-cyclic olefins, HBr addition is not stereospecific except at very low temperatures, when bond rotation is appreciably slower and when the reversibility of attack by Br· is minimized.

11.4.2 *Addition of halogens*

The free-radical addition of halogen to olefins again involves attack by X· initially. This process is reversible for I_2 and Br_2 attack at room temperature (as shown by the *cis-trans* isomerization of olefins in the presence of traces of these halogens), but is only appreciably so for Cl_2 at temperatures above 200 °C. At increasing temperatures, however, hydrogen abstraction becomes more important, since the simple adducts are thermally unstable (i.e. the addition is reversible).

$$-CH_2-\overset{|}{C}=C\langle + X_2 \rightleftharpoons -CH_2-\overset{|}{C}(X)-C(X)\langle$$

$$\downarrow -HX$$

$$-CH(X)-\overset{|}{C}=C\langle$$

Since halogen-atom addition is reversible, allylic halides may give rearranged products through the formation of a radical such as $-CHBr-\dot{C}H-CHBr-$ which may lose either halogen atom as Br· and, in consequence, generate either a new allylic halide or the starting material.

$$R-CH=CH-CHBr- + Br\cdot \longrightarrow R-CHBr-\dot{C}H-CHBr-$$

$$R-CH=CH-CHBr- + Br\cdot \quad or \quad R-CHBr-CH=CH- + Br\cdot$$

11.5 Arylation by benzoyl peroxides

The decomposition of peroxides in most organic media can involve two routes. A so-called 'spontaneous' decomposition can provide two acyloxy radicals by homolysis of the O—O bond,

$$RCOOOCOR \rightarrow 2RCOO\cdot,$$

while an 'induced' decomposition, which involves attack by a radical already in solution, provides one acyloxy group,

$$M\cdot + RCOOOCOR \rightarrow RCOOM + RCOO\cdot.$$

The fate of the acyloxy radicals then depends upon reaction conditions. Carbon dioxide may be lost to give the reactive alkyl or aryl ($R\cdot$) radicals, or the acyloxy radical itself may initiate some other process (e.g. polymerization of an olefin).

In most aromatic solvents the decomposition of benzoyl peroxides (e.g. $(PhCO.O)_2$) follows the kinetic form

$$Rate = k_s[PhCOOOCOPh] + k_i[PhCOOOCOPh]^{3/2},$$

where k_s and k_i refer to the spontaneous and induced decomposition respectively. The numerical value of k_i does not alter appreciably over a range of relatively inert solvents, showing that induced decomposition is not critically dependent upon radicals derived from the solvent; in more suitable conditions, when the solvent may provide a source of radicals (e.g. acetic acid, aliphatic ethers, and some alcohols), the values of k_i increase considerably.

In this way, we can imagine a number of substitution reactions by the aryl radicals upon the aromatic solvent, exemplified by the process

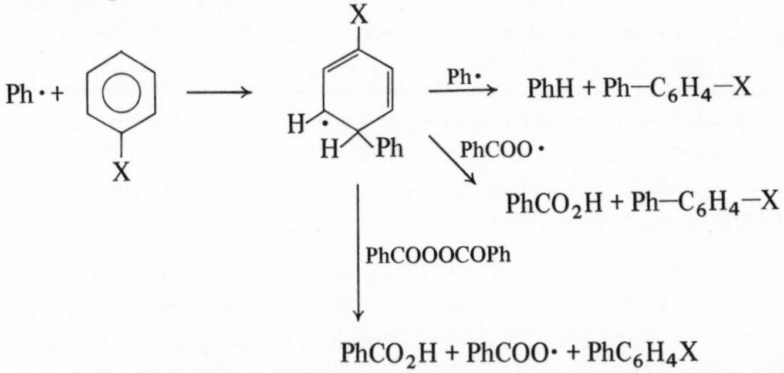

when the aromatization of the initially formed radical (the so-called 'sigma' complex) is achieved either by reaction with another radical or by attack upon benzoyl peroxide. All of these processes allow hydrogen abstraction. A second mode of reaction would involve the dimerization of the sigma complex, forming tetrahydroquaterphenyls:

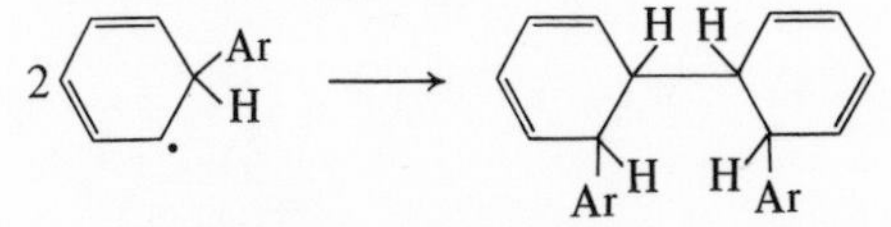

These species have not been directly isolated, for during distillation or other treatment they are oxidized to form the fully aromatic species; the production of these quaterphenyls may be taken as evidence of an intermediate 'sigma' radical. Quaterphenyls might also result from further arylation of the initially formed biphenyl products. This can be shown not to be the case (i) by using

substituted aroyl peroxides and (ii) by the unexpectedly high concentration
of quaterphenyls in the polyaryl fractions.

Problems

11.1   When *N*-nitrosoacetanilide decomposes in aromatic solvents, biphenyls are formed
from attack upon the solvent. The ratio of *ortho-*, *meta-* and *para-*substituted
products is similar to that found when benzoyl peroxide is decomposed in the
same solvent. What is the apparent attacking group? The rates of decomposition
of *N*-nitrosoacetanilide in various aromatic solvents are very similar to each other,
although the products differ; similarly, the rate of decomposition of the reagent
is unaffected by added β-naphthol, although some of the product is diverted to
the corresponding azo-dye. Discuss the significance of these results. (See Williams,
1960; Cadogan, 1970.)

11.2   Suggest reasons for the following:

(a) The exchange of halogen between radioactive I_2 and $PhCH_2I$ occurs readily
under irradiation, and proceeds more easily in carbon tetrachloride than in
hexachloro-1,3-butadiene.

(b) Cyclopentene reacts with dinitrogen tetroxide to give 2-nitrocyclopentanol
nitrite; in the presence of $BrCCl_3$, 2-nitrocyclopentyl bromide is formed.

(c) The formation of ArCl from ArN_2^+ by cuprous chloride in HCl solutions
apparently involves a radical reaction between the diazonium ion and $CuCl_2^-$, when
nitrogen, cupric chloride, and Ar· are formed. How does this explain the isolation
of $ArCH_2CHClCN$ when the reaction is carried out in the presence of acrylonitrile?

(d) When propionyl peroxide (EtCOOOCOEt) is decomposed in hydrocarbon
solvents, one of the products is *n*-butane. The yield of butane does *not* depend
upon the initial concentration of peroxide. How does this suggest that the ethyl
radicals are not 'free' (i.e. do not have discrete existence in the solvent for any
length of time)? Quinone, a good scavenger of alkyl radicals, is not attacked in
such solutions of decomposing peroxide. Is this consistent with a restricted life of
the ethyl radical?

Chapter 12
Carbanion Chemistry

2.1   **Carbanions and their reactions**

We have already seen that the carbon–hydrogen bond is not a particularly polar one, and that the acidity of normal alkanes is remarkably small ($K \sim 10^{-40} - 10^{-50}$). Substituents attached to the carbon atom could, by withdrawing electrons, increase the polarity or acidity of the C–H bond and might even allow the resulting carbanion to have some stability in aqueous media. For instance, three halogen atoms (e.g. $CHCl_3$) sufficiently increase the polarity of the C–H bond in the haloforms that the trihalogenomethide ion (CX_3^-) can exist as an intermediate in the base-catalysed hydrolysis of chloroform or bromoform (equation **12.1**).

$$HCX_3 \rightleftharpoons H^+ + CX_3^- \qquad\qquad\qquad \textbf{12.1}$$

In cases where the resulting anion can be stabilized by resonance, remarkable increases in acidity result upon substituting hydrogen atoms in methane. Thus, CH_3NO_2 ($K_a^* = 6 \times 10^{-11}$) and ethyl acetoacetate ($K_a^* = 2 \times 10^{-11}$) show appreciable acidity in water; the effect is increased, although not proportionately, by further substitution. Dinitromethane therefore shows $K_a^* = 2.5 \times 10^{-4}$, and both $HC(NO_2)_3$ and $HC(SO_2Me)_3$ show $K_a^* = 0.1$.

While acidifying the basic solutions of these keto-esters and hydrocarbons usually provides the original acid again, the nitroalkane anion (e.g. $[CH_2NO_2]^-$) gives a much more acidic *aci*-nitroalkane (12.1); this species rearranges to form the original species (12.2) quite slowly. These two *tautomers* interconvert through a *prototropic* shift, in the same way as the keto–enol tautomerism occurs in the

$$CH_2=\overset{+}{N}\!\!\big\langle{}^{OH}_{O^-} \qquad\qquad CH_3-\overset{+}{N}\!\!\big\langle{}^{O}_{O^-} \quad \text{(H moves)}$$

(12.1) *aci*-form   (12.2)

$$
\begin{array}{ll}
CH_3-C=O & CH_3-C-OH \\
\quad\ \ | & \qquad \| \\
\quad\ \ CH_2 & \qquad CH \quad \text{(H moves)} \\
\quad\ \ | & \qquad | \\
\quad\ \ CO_2Et & \qquad CO_2Et
\end{array}
$$

(12.3) *keto*-form   (12.4) *enol*-form

* These are apparent dissociation constants, without any allowance for the existence of tautomers, one of which would be the true acid (e.g. keto–enol pairs).

β-keto-esters or the β-diketones (12.3, 12.4). The two tautomers are different isomers coexisting in the one liquid (cf. resonance forms!). It is possible to separate the two species by purely physical methods (e.g. aseptic fractional distillation) although they are in equilibrium under normal conditions.

12.2   Hydrocarbons of high acidity

1,3-Cyclopentadiene (12.5) shows unusually high acidity, and is able to form the anion in liquid ammonia, or even (in significant amounts) in solutions of ethoxide ion. The high stability of the anion is associated with the formation of an aromatic sextet of electrons; the carbanion has some degree of aromatic stability. Less stabilization of the anion is found in indene (benzocyclopentadiene, 12.6) and in fluorene (dibenzocyclopentadiene 12.7) and is shown by the corresponding decrease in the acidities of the three hydrocarbons.

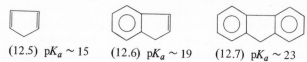

(12.5) $pK_a \sim 15$ (12.6) $pK_a \sim 19$ (12.7) $pK_a \sim 23$

None of these hydrocarbons show appreciable acidity in water, and their anions (conjugate bases) are completely hydrolysed by this solvent. In contrast, fluoradene (indeno-1,2,3-[jk]fluorene) can be extracted from benzene solution by *dilute, aqueous* sodium hydroxide, and is undoubtedly one of the most acidic simple hydrocarbons.

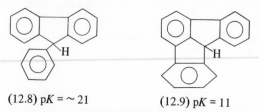

(12.8) pK = ∼ 21   (12.9) pK = 11

12.3   Stabilities and properties of carbanions

The carbonium ions, R_3C^+, have been shown to prefer a planar configuration around C_α. The fact that ketones such as the diketone (12.10)

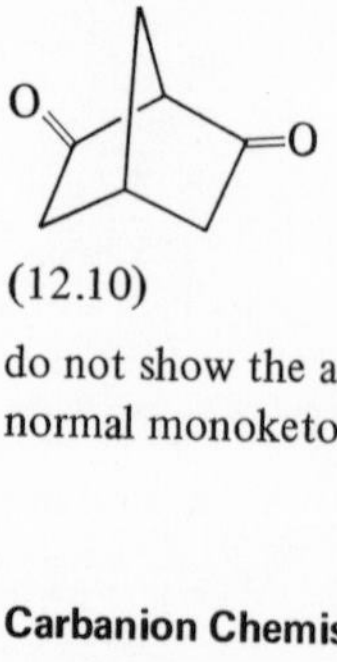

O ⟍ ⟋ O

(12.10)

do not show the acidity associated with a β-diketone, but rather that due to a normal monoketone structure, $-CH_2-CO-$, may be used as evidence that carbanions also require coplanarity of the α-carbon atom. In the same way, the lower acidity of triptycene in comparison with triphenylmethane may be interpreted in terms of the lack of planarity of the carbanion and the resulting loss of stabilization. The apparent planarity of the carbanion seems not to be an initial result of acquiring the R_3C^- structure, but an essential for its stabilization. In all cases in which such ions can be studied, the most stable are those in which the carbanionic centre is part of a conjugated system, whose resonance stabilization requires coplanarity of the atoms. The only instances in which the ionic centre is not part of such a stabilizing system (e.g. sodium or potassium alkyls) are highly reactive species in which R_3C^-, if it exists at all, is stabilized by crystal-lattice forces. In solution, such alkyls immediately attack the solvent.

Allowing such limits to be placed upon our knowledge of carbanions, could we predict any types of reaction? There would be expected to be nucleophilic addition processes across C=O and C=C structures, and the effects of substituents upon the ease of such additions would be opposite to those involving electrophilic attack (e.g. halogen addition). There are also some types of reaction which are not so easily predicted.

2.4 Base-catalysed reactions

2.4.1 *Base-catalysed halogenation of ketones*

The chlorination, bromination or iodination of ketones occurs rapidly to give the substitution product,

$$RCH_2COR' \xrightarrow[-HX]{X_2} RCHXCOR'.$$

Since substitution occurs at C_α, it is evident that the reaction is mainly confined to enolizable ketones. The rate of the reaction, however, is independent of the nature or the concentration of the halogen,

Rate = k[ketone] [OH$^-$].

Presumably the slow stage of the reaction involves base and ketone alone; the halogen is introduced in a subsequent and rapid process. The most simple mechanism fulfilling these requirements would involve a slow formation of the ketone carbanion, followed by rapid electrophilic attack by halogen:

$$-CH_2CO- \underset{slow}{\overset{base}{\rightleftharpoons}} -\bar{C}H-CO-,$$

$$-\bar{C}H-CO- + X_2 \xrightarrow{fast} -CHX-CO- + X^-.$$

If this deprotonation is the slow stage of the process, we might expect that the subsequent reaction with *any* electrophile could be a fast process. In keeping with this, the rates of bromination and of racemization of the optically active ketone Et(Me)CHCOPh are the same in aqueous acetic acid, and these rates are the same as the rate of uptake of deuterium from a deuterated solvent. In these instances the electrophiles are Br_2, H_2O and D_2O respectively.

12.4.2 *Base-catalysed aldol condensation*

The dimerization of acetaldehyde is catalysed by both acids and bases. Such an 'aldol' condensation is general to aldehydes, and to most ketones, having available α-hydrogen atoms, e.g.

$$RCH_2COR' + R''CHO \rightarrow \underset{\underset{CH(OH)R''}{|}}{RCHCOR'}.$$

We can imagine these reactions, and the corresponding Claisen condensations (section 12.4.3), as involving nucleophilic attack by a carbanion upon the carbonyl system. The base now allows the formation of equilibrium quantities of the carbanion which may subsequently attack C=O. The resulting anion will then be in equilibrium with its conjugate acid.

$$RCH_2COR' + base \rightleftharpoons R\bar{C}HCOR' + [base{-}H]^+$$

$$R\bar{C}HCOR' + R''CHO \rightleftharpoons \underset{\underset{CH(O^-)R''}{|}}{RCHCOR'}$$

$$\underset{\underset{CH(O^-)R''}{|}}{RCHCOR'} + [base{-}H]^+ \rightleftharpoons \underset{\underset{CH(OH)R''}{|}}{RCHCOR'} + base$$

The first and the third processes will presumably be equilibria; the second stage is presumably so, since aldol condensations are found to be reversible unless the equilibria are affected by dehydration of the aldol (irreversibly). However, we can predict little else about the relative rates of the processes, although the third process will probably be fast, since it involves simple proton transfer and does not require any rearrangement or isomerization to occur.

Two extremes can be considered. If the rate of formation of the carbanion were sufficiently slow to be the controlling step, we would expect a rate equation

Rate = k[base] [aldehyde]

for the dimerization of acetaldehyde, since all other stages in the condensation will be fast, and so kinetically not significant. Such a kinetic form is found in hydroxide-ion-catalysed condensations with high initial concentrations of acetaldehyde. Under these conditions, since the deprotonation stage will be virtually irreversible, no isotope exchange should occur between the aldehyde and the solvent. No such exchange is found to occur in either the starting material or the aldol product when the condensation is carried out in D_2O.* Hence the slow stage of the reaction would seem to be the first step in which the carbanion is formed; its subsequent reactions are fast sequences which are apparently irreversible under the reaction conditions.

---

* This applies, of course, to C–D bond formation only, and not to O–D bond formation, which proceeds more rapidly.

The other extreme is found in the condensation of malonic ester with formaldehyde, or the formation of benzalacetophenone,

$$CH_2O + \bar{C}H(CO_2Et)_2 \xrightarrow{OH^-} HO-CH_2-CH(CO_2Et)_2,$$

$$PhCHO + H_3CCOPh \xrightarrow{OEt^-} PhCH=CHCOPh.$$

The kinetic form becomes rather complex, but it may be rationalized in terms of a fast, pre-equilibrium in which the carbanion is formed, e.g.

$$CH_2(CO_2Et)_2 \underset{}{\overset{base}{\rightleftharpoons}} \bar{C}H(CO_2Et)_2,$$

followed by a slow addition of the anion to the carbonyl group, e.g.

$$H_2C=O + \bar{C}H(CO_2Et)_2 \longrightarrow \underset{\underset{CH_2O^-}{|}}{CH(CO_2Et)_2}$$

The rate of the reaction now depends upon the concentrations of both aldehyde and carbanion, and hence is expressed by

$$Rate = kK_{BH+} \frac{[\text{carbonyl species}]\,[RCOCH_2R']\,[\text{base}]}{[\text{base}-H^+]},$$

where K_{BH+} refers to the proton affinity of the base B, and the species $RCOCH_2R'$ provides the carbanion. If B is the conjugate base of the solvent (e.g. OEt^- in EtOH) then the term $[Base.H^+]$ vanishes and appears in the overall constant. Such a kinetic form, besides being found in the two examples quoted, also appears in the hydroxide-ion-catalysed dimerization of acetaldehyde at low concentrations,

$$Rate = k[OH^-]\,[CH_3CHO]^2.$$

Since the carbanion is formed reversibly, deuterium exchange with the solvent can be observed.

2.4.3   Claisen condensation

The aldol condensation involves carbanionic attack of C=O in aldehydes and ketones. The same process, involving the carboxylic esters, is the basis of the Claisen condensation. Carbanions apparently are present in small amount in mixtures of strong bases and esters, ketones, or aldehydes with α-hydrogen atoms; hydrogen exchange occurs between solvent (e.g. EtOD) and the α-carbon atoms under these conditions, and optically active esters (e.g. Et.CH(Me).CO$_2$Et) are racemized, presumably via a planar carbanion, on treatment with OEt$^-$. The Claisen condensation may therefore follow a mechanism similar to the hydrolysis of esters, i.e.

$$\text{base} + RCH_2CO_2Et \rightleftharpoons R\bar{C}HCO_2Et + [\text{base}-H]^+,$$

$$RCH_2\underset{\underset{OEt}{|}}{C}{=}O + \bar{C}H(R)CO_2Et \rightleftharpoons RCH_2CCHRCOEt$$
$$\underset{EtO \quad O^-}{}$$

$$RCH_2\underset{\underset{OEt}{|} \quad O^-}{C}CHRCO_2Et \rightleftharpoons RCH_2COCHRCO_2Et + OEt^-$$

$$RCH_2COCHRCO_2Et + OEt^- \rightleftharpoons RCH_2CO\bar{C}RCO_2Et + EtOH.$$

The reaction may be brought to completion only by removing or stabilizing the product. This is usually done by removing the more acidic β-ketoester as its salt. When dialkyl-acetic esters (R_2CHCO_2Et) are involved, the β-ketoester has no α-hydrogen atoms. The condensation does not then provide a good yield of β-ketoester until a sufficiently strong base is used as catalyst for the stabilization of the anion $R_2\bar{C}COCR_2CO_2Et$, i.e. for the successful removal of a γ-hydrogen atom.

12.4.4 *Michael reaction*

A vast number of addition reactions fall under this heading. In the broadest possible sense, they involve the addition of a carbanion across a carbon–carbon double bond, usually one which has been activated by being in conjugation with C=O (e.g. α,β-unsaturated esters and aldehydes) or with C≡N (e.g. acrylonitrile). An example is the condensation of diethyl malonate with ethyl crotonate:

$$CH_2(CO_2Et)_2 \xrightleftharpoons{OEt^-} \bar{C}H(CO_2Et)_2 + EtOH,$$

$$CH_3CH{=}CHCO_2Et + \bar{C}H(CO_2Et)_2 \rightleftharpoons CH_3\underset{\underset{CH(CO_2Et)_2}{|}}{\overset{\frown}{C}HCHCO_2Et}$$

$$CH_3\underset{\underset{CH(CO_2Et)_2}{|}}{\overset{\frown}{C}HCHCO_2Et} \xrightleftharpoons[OEt^-]{EtOH} CH_3\underset{\underset{CH(CO_2Et)_2}{|}}{C}HCH_2CO_2Et \xrightleftharpoons[EtOH]{OEt^-} CH_3\underset{\underset{\bar{C}(CO_2Et)_2}{|}}{C}HCH_2CO_2Et$$

Since the product is a weaker acid than the starting material (H replaced by R— attached to the active methylene group) the reaction will be hindered by excessive amounts of base (cf. Claisen condensation) which will stabilize the reagent at the expense of the products. As with electrophilic addition (acid-catalysed hydration, or addition of halogen), conjugated dienoic esters ($RCH{=}CHCH{=}CHCO_2Et$) and their vinylogues* will undergo 1,4- or 1,6-addition as well as 1,2-addition; the intermediate carbanion would have charge spread over five or seven atoms instead of the three depicted in the reaction scheme.

* $R(CH{=}CH)_nCO_2Et$, containing the conjugated π-electron system $-(CH{=}CH)_n-CO-$.

Acid-catalysed condensation reactions

While it is not strictly relevant to a study of carbanion chemistry, the fact that both the halogenation of ketones and the aldol condensation are also subject to *acid*-catalysis should be mentioned. The halogenation of enolizable ketones is subject to general acid-catalysis, and, as with the base-catalysed process, the rate of halogenation (regardless of halogen), of racemization of resolved ketones with optically active asymmetric α-carbon atoms, and of deuterium exchange is the same. Since only the ketone and the acid are involved in the rate-determining stage,

Rate = k[ketone] [acid],

it is the formation of some intermediate which is the common, slow process.

This intermediate might be the ketone–acid complex $[R_2CO \cdots HX]$. or even the protonated ketone itself. However, the rate of acid-catalysed oxygen exchange between $H_2{}^{18}O$ and the ketone is much greater than the rate of bromination under identical conditions of acidity. Oxygen exchange might involve a mechanism such as

$$R_2C=O + HX \rightleftharpoons R_2C=O \cdots HX \rightleftharpoons \underset{\underset{OH_2}{|}}{R_2C}=O \cdots HX \rightleftharpoons \underset{\underset{OH}{|}}{R_2C}\text{--OH} \rightleftharpoons \underset{\underset{O}{\|}}{R_2C} + OH_2$$

in which protonation (or complexing with HX) increases the electrophilicity of the carbonyl carbon atom towards attack by solvent, followed by prototropy and the elimination of the original carbonyl oxygen atom as water. Since the whole process is more rapid than bromination, the formation of the ketone–acid complex cannot be the slow stage in halogenation. However, we could imagine that enol formation might be a suitably slow process. Enolization could occur by the sequence

$$\underset{\underset{R'}{|}}{\underset{CH_2}{|}}R\text{--}C=O \cdots HX \xrightarrow{\text{base}} \underset{\underset{R'}{|}}{\underset{CH \cdots H \cdots \text{Base}}{|}}R\text{--}C\text{:}\cdots\text{:}O \cdots H \cdots X \quad \underset{\underset{R'}{|}}{\underset{CH}{\|}}R\text{--}C\text{--OH} + X^- + \text{Base--H}^+$$

in which the nucleophile is attacking the α-carbon atom and not the more polarized carbonyl carbon atom. The nucleophilicities of the two reagents ($H_2{}^{18}O$ and a weak base which might well be the solvent) would be similar, and the lesser electronegativity of the α-carbon atom would explain the lower rate of attack:

$$R\text{--}\underset{\underset{\overset{\displaystyle|}{H}}{|}}{\overset{\overset{\displaystyle\delta\delta +}{H}}{C}}\text{--}\overset{\delta +}{C}\underset{{}^{18}OH_2}{\text{--}O\cdots H\text{--}X.}$$

The enol would then undergo a very rapid reaction with halogen, or with deuterated solvent (cf. vinyl ethers and their rates of halogenation). Since a variety of acids could act as HX (since the process is subject to general acid catalysis) and since a variety of bases could assist in the removal of an α-hydrogen atom (e.g. the conjugate base X^-, and possibly the solvent), the kinetic form of the reaction is, in principle, complex. This complexity is rarely fully appreciated, since at any particular acidity there may be significant contributions by only two or three of the terms (due to the vanishingly small concentrations of some other of the reagents), and of these one is often much greater than the other two terms. For instance, in the bromination of acetone in aqueous acetic acid buffers, the contribution by terms involving the base OH^- would be very small (concentration effect). But the extent of the catalysis by H_3O^+, as measured by the numerical value of $k_{H_3O^+}$, might be much larger than that due to HOAc (measured by k_{HOAc}) and hence it would not be particularly difficult to find a situation in which almost all of the observed kinetic results could be ascribed to H_3O^+ catalysis alone.

Similar effects are found in the acid-catalysed aldol condensation reaction. Here the rate of the reaction appears to depend upon Hammett acidity functions in acetic acid solution, indicating the intermediacy of $[R_2COH]^+$; a consistent mechanism would involve the rate-determining attack of the ketone conjugate acid upon the C=C structure of the enol.

Problems

12.1   Suggest mechanisms for the following reactions:

(a) $BrCH_2CH_2CH_2CN + NaNH_2 \rightarrow$;

(b) $PhCH=C(CO_2Et)_2 + ArMgX \rightarrow Ph(Ar)CHCH(CO_2Et)_2$;

(c) $PhCH=C(CN)CO_2^- + CN^- \rightarrow PhCH(CN)\underset{\underset{H}{|}}{C}(CN)CO_2^-$

(d) $(RCH_2CO)_2O + PhCHO \xrightarrow{\text{base}} PhCH=C(R)CO_2H$;

(e) 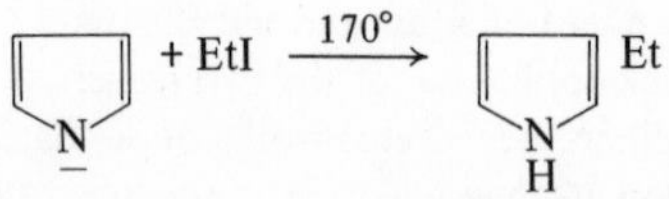 $+ EtI \xrightarrow{170°}$ Et

(f) 5-methoxyindene $\xrightarrow[\text{EtOH}]{\text{NaOEt}}$ 6-methoxyindene.

2 Explain the following:

(a) Tri-*o*-tolyl methane, on treatment with strong base ($KNEt_2$) followed by CO_2, gives a phenyl-acetic acid derivative (indicating metallation of $-CH_3$) whereas tri-*p*-tolyl methane under the same conditions gives tris(*p*-tolyl)acetic acid.

(b) Attempted hydrolysis of $BrCH(CO_2Et)_2$ by aqueous HBr gives free halogen.

(c) When the iodoform reaction is carried out using acetone in D_2O, the iodoform, but not the acetic acid formed, is labelled.

3 2-Chlorocyclohexanone reacts with hydroxide ion to give cyclopentanoic acid. Devise a possible mechanism.

(a) When labelled (C-2) chlorocyclohexanone is used, half of the labelling appears at the C-2 carbon atom, and half at the C-1 carbon atom of the cyclopentane-1-carboxylic acid. Does your mechanism allow for this?

(b) One explanation of this isotope 'scrambling' is the formation of a cyclopropanone structure. Suggest a way in which you would test this hypothesis.

(c) 1-Benzoyl-1-chlorocyclohexanone also rearranges under equally mild conditions to give 1-phenyl-cyclohexanoic acid. Can this be accommodated in the proposed mechanism?

(N.B. In fact, two mechanisms are necessary, depending upon the structure of the halogenoketone. The Favorskii rearrangement has been a source of much controversy.)

References

Arnold, R. T., and Buckley, J. S. (1949), *Journal of the American Chemical Society*, vol. 71, p. 1781.

v. Auwers, K. (1923), *Berichte des deutschen chemischen Gesellschaft*, vol. 56, p. 715.

Bartlett, P. D., and Brown, R. F. (1940), *Journal of the American Chemical Society*, vol. 62, p. 2927.

Brown, H. C., (1956), *Journal of the Chemical Society*, p. 1248.

Brown, H. C., and McGaryn, C. W. (1955), *Journal of the American Chemical Society*, vol. 77, p. 2306.

Cadogan, J. I. (1970), 'Essays in free-radical chemistry', *Chemistry Society Special Publications*, vol. 24, p. 71.

Cristol, S. J. (1947), *Journal of the American Chemical Society*, vol. 69, p. 338.

Cristol, S. J. (1949), *Journal of the American Chemical Society*, vol. 71, p. 1894.

Cristol, S. J. (1951), *Journal of the American Chemical Society*, vol. 73, p. 674.

Cristol, S. J. (1953), *Journal of the American Chemical Society*, vol. 75, p. 2647.

Cristol, S. J., and Norris, W. P. (1953), *Journal of the American Chemical Society*, vol. 75, p. 2645.

Deno, N. C., *et al.* (1962), *Journal of the American Chemical Society*, vol. 84, p. 1498.

Deno, N. C., *et al.*, (1963), *Journal of the American Chemical Society*, vol. 85, p. 2991.

Eberson, L., Winstein, S., *et al.* (1965), *Journal of the American Chemical Society*, vol. 87, pp. 3504 and 3506.

Grovenstein, E., and Lee, D. E. (1954), *Journal of the American Chemical Society*, vol. 75, p. 2639.

Hauser, C. R., Le Maistre, J. W., and Reinsford, A. E. (1935), *Journal of the American Chemical Society*, vol. 57, p. 1056.

Hofmann, P. S., *et al.* (1960), *Recueil des Travaux Chimiques des Pays-Bas*, vol. 79, p. 790.

Lynch, B. M., and Pausacker, K. (1955), *Journal of the Chemical Society*, p. 1525.

de la Mare, P. B. D., *et al.* (1958), *Chemistry and Industry*, p. 1806.

Nightingale, D. V. (1947), *Chemical Reviews*, vol. 40, p. 117.

Ogata, Y., and Okano, M. (1956), *Journal of the American Chemical Society*, vol. 78, p. 5423.

Pearson, R. G., *et al.*, (1951), *Journal of the American Chemical Society,* vol. 73, pp. 923 and 931.

Puterbaugh, W. H., and Newman, M. S. (1959), *Journal of the American Chemical Society,* vol. 81, p. 1611.

Roberts, I., and Urey, H. C. (1938), *Journal of the American Chemical Society,* vol. 60, p. 880.

Roberts, J. D., *et al.* (1951), *Journal of the American Chemical Society,* vol. 73, p. 2509.

Roberts, J. D., *et al.* (1959), *Journal of the American Chemical Society,* vol. 86, p. 3773.

Westheimer, F. H. (1936), *Journal of the American Chemical Society,* vol. 58, p. 2209.

Westheimer, F. H., Segel, E., and Schram, R. (1947), *Journal of the American Chemical Society,* vol. 69, p. 773.

Williams, G. H. (1960), *Homolytic Aromatic Substitution*, Pergamon.

Further Reading

This list is of principal references and review articles which should be useful in supplementary reading. There are a number of texts dealing with physical organic chemistry which may be consulted; it is wise in some cases to ensure, by cross-reference to another text, whether some mechanisms and suggestions are generally accepted or are an expression of the author's opinion. In the fundamental work of the first three chapters, however, there is very little divergence between texts.

The following texts are particularly relevant. *Mechanism and Structure in Organic Chemistry* by E. S. Gould (Holt-Dryden, 1960), and *Physical Organic Chemistry* by J. Hine (McGraw-Hill, 1962, 2nd edn) have been praised in the Preface. *A Guidebook to Mechanism in Organic Chemistry* by P. Sykes (Longman, 1970, 3rd edn) may also be useful, and *Advanced Organic Chemistry* by J. March (McGraw-Hill, 1968) deserves especial mention for its combination of classical organic chemistry with mechanistic studies.

Chapter 1

R. J. Gillespie, *Journal of Chemical Education*, vol. 40 (1963), p. 295. (Hybridization.)

L. Pauling, *The Nature of the Chemical Bond*, 3rd edn, Cornell University Press, 1960. (Bonding and electronegativity.)

A. Streitwieser, *Molecular Orbital Theory for Organic Chemists,* Wiley, 1961. (Atomic and molecular orbital theory.)

G. Wheland, *Resonance in Organic Chemistry*, Wiley, 1955. (Resonance and delocalization; see also Streitwieser, *loc. cit.*)

Chapter 2

E. M. Arnett, in S. G. Cohen (ed.), *Progress in Physical Organic Chemistry*, Interscience, 1963, vol. 1, p. 223. (Acids and bases.)

R. P. Bell, *Quarterly Reviews*, vol. 1 (1947), p. 113. (Acids and bases.)

K. Bowden, *Chemical Reviews*, vol. 66 (1966), p. 119. (Acidity functions.)

H. C. Brown, *et al.*, *Journal of the American Chemical Society*, vol. 79 (1959), pp. 1906, 1909, 1913. (Hammett linear free-energy relationships.)

H. C. Brown, in G. W. Gray (ed.), *Steric Effects in Conjugated Systems*, Butterworths, 1958, p. 100. (Hammett linear free-energy relationships.)

J. F. Bunnett and F. Olsen, *Chemical Communications*, 1965, p. 601. (Acidity functions.)

J. F. Bunnett and F. Olsen, *Canadian Journal of Chemistry*, vol. 44 (1966), p. 1897. (Acidity functions.)

C. J. Collins, *Advances in Physical Organic Chemistry*, vol. 2 (1964), p. 3. (Isotopic labelling.)

S. L. Freiss, E. Lewis, and A. Weissberger (eds.), 'Investigation of Rates and Mechanisms of Reactions', Parts 1 and 2, *Techniques in Organic Chemistry*, Interscience, 1963, vol. 8.

A. A. Frost and R. G. Pearson, *Kinetics and Mechanism*, 2nd edn, Wiley, 1961. (Kinetic methods.)

H. H. Jaffe, *Chemical Reviews*, vol. 53 (1953), p. 191. (Hammett linear free-energy relationships.)

L. Melander, *Isotope Effects on Reaction Rates*, Ronald Press, 1960. (Isotope effects.)

B. Wepster, *et al.*, *Recueil des Travaux Chimiques des Pays-Bas*, vol. 78 (1959), p. 815.

Chapter 3

P. D. Bartlett, *Non-classical Ions*, Benjamin, 1965.

D. Bethell and V. Gold, Carbonium Ions, Academic Press, 1967.

Chapter 4

H. C. Brown, *Journal of the Chemical Society*, 1956, p. 1248. (Steric effects in S_N2 reactions.)

C. A. Bunton, *Nucleophilic Substitution at a Saturated Carbon Atom*, Elsevier, 1963.

B. C. Capon, *Quarterly Reviews*, vol. 18 (1964), p. 45. (Neighbouring-group participation.)

C. K. Ingold, *Structure and Mechanism in Organic Chemistry*, 2nd edn, Bell, 1969. (S_N1 and S_N2 mechanisms.)

Chapter 5

D. V. Banthorpe, *Elimination Reactions*, Elsevier, 1963.

J. F. Bunnett, *Angewandte Chemie* (International Edition in English), vol. 1 (1962), p. 225.

W. H. Saunders, in S. Patai (ed.), *The Chemistry of Alkenes*, Interscience, 1964.

Chapter 6

E. **Berliner**, in S. G. Cohen (ed.), *Progress in Physical Organic Chemistry*, Interscience, 1964, vol. 2, p. 253.

P. B. D. **de la Mare** and J. H. **Ridd**, *Aromatic Substitution–Nitration and Halogenation*, Butterworths and Academic Press, 1959.

R. O. C. **Norman** and R. **Taylor**, *Electrophilic Substitution in Benzenoid Compounds*, Elsevier, 1965.

G. **Olah** (ed.), *Friedel-Crafts and Related Reactions*, Interscience, 1963.

H. **Zollinger**, *Azo and Diazo Chemistry*, Interscience, 1961.

H. **Zollinger**, *Advances in Physical Organic Chemistry*, vol. 2 (1964), p. 163.

Chapter 7

J. F. **Bunnett**, *Quarterly Reviews,* vol. 12, 1958, p. 1.

J. F. **Bunnett**, *Journal of Chemical Education*, vol. 38 (1961), p. 278.

J. F. **Bunnett** and R. E. **Zahler**, *Chemical Reviews*, vol. 49 (1951), p. 273.

J. **Miller**, *Aromatic Nucleophilic Substitution*, Elsevier, 1969.

K. **Wittig**, *Pure and Applied Chemistry*, vol. 7 (1963), p. 173.

K. **Wittig**, *Angewandte Chemie* (International Edition in English), vol. 4 (1965), p. 731.

Chapter 8

R. P. **Bell**, *Advances in Physical Organic Chemistry*, vol. 4 (1966), p. 1.

M. J. S. **Dewar**, *Angewandte Chemie* (International Edition in English), vol. 3 (1964), p. 245.

J. **Jencks**, in S. A. Cohen (ed.), *Progress in Physical Organic Chemistry*, Interscience, 1964, vol. 2, p. 63.

P. B. D. **de la Mare**, *Quarterly Reviews*, vol. 3 (1949), p. 126.

P. B. D. **de la Mare** and R. **Bolton**, *Electrophilic Additions to Unsaturated Systems*, Elsevier, 1966.

S. **Patai** and Z. **Rappoport**, in S. **Patai** (ed.), *The Chemistry of Alkenes*, Interscience, 1964, p. 469.

Chapter 9

C. K. **Ingold**, *Structure and Mechanisms in Organic Chemistry*, 2nd edn, Bell, 1969.

D. P. N. **Satchell**, *Quarterly Reviews*, vol. 17 (1963), p. 160.

N. O. V. **Sonntag**, *Chemical Reviews*, vol. 52 (1953), p. 237.

Chapter 10

D. V. Banthorpe, E. D. Hughes and C. K. Ingold, *Journal of the Chemical Society*, 1964, p. 2864.

C. J. Collins, *Quarterly Reviews*, vol. 14 (1960), p. 357.

M. J. S. Dewar, in P. de Mayo (ed.), *Molecular Rearrangements*, Interscience, 1963, vol. 1, p. 295.

P. de Mayo, *Molecular Rearrangements*, Interscience, 1963. (General.)

H. Shine, *Aromatic Rearrangements*, Elsevier, 1967.

Chapter 11

J. I. G. Cadogan and M. J. Perkins, in S. Patai (ed.), *The Chemistry of Alkenes*, Interscience, 1964, p. 585.

W. A. Pryor, *Free Radicals*, McGraw-Hill, 1966,

C. Walling, *Free Radicals in Solution*, Wiley, 1957.

G. H. Williams, *Homolytic Aromatic Substitution,* Pergamon, 1960.

Chapter 12

D. J. Cram, *Fundamentals of Carbanion Chemistry*, Academic Press, 1965.

C. R. Hauser, and B. E. Hudson, *Organic Reactions*, vol. 1 (1942), p. 266.

H. A. House, *Modern Synthetic Reactions*, Benjamin, 1965.

W. S. Johnson and G. H. Daub, *Organic Reactions*, vol. 6 (1951), p. 1.

R. P. Bell, *Quarterly Reviews*, vol. 1 (1947), p. 113.

C. K. Ingold, *Structure and Mechanisms in Organic Chemistry*, Bell, 1969.

A. Streitwieser, *Molecular Orbital Theory for Organic Chemists*, Wiley, 1961.

Aids to Problems

In this section the complete solutions to each set of problems are not given. The comments given here are intended to supplement the text so that the answer should become apparent. In some cases, however, where no aid is given, the best answer is to refer the reader to the reference given at the end of the problem, when the logic behind the proposed mechanism may be more clearly appreciated.

Chapter 1

1.1 The nucleus has a very small volume. If the electron is confined to these limits, what can we deduce about its energy? (Heisenberg principle.)

1.2 What are the electronegativities of the atoms?

1.3 Can the resulting anions be stabilized, either by inductive effects or by resonance?

1.4 How will electron-withdrawal affect the acidity of phenol? Which of the components has the higher I effect? Which has the greater M effect? Which therefore has the larger aggregate effect? (Remember the variation of efficient orbital overlap with size.)

1.5 Decrease (why?).

1.6 Attack by an electrophile will presumably involve the π-clouds, so that attack is directional (in one plane) in olefins, but possible over a far wider angle with acetylenes, which are linear.

1.7 See page 21.

1.8 Might not a dipole, by changing the electron density and hence electronegativity of the atom, cause some perturbation of a π-system involving one of the atoms of the dipole?

Chapter 2

2.1 (a) Loss of Br_2 (spectrophotometrically, or by quenching in KI and titrating iodine); loss of diene (u.v., i.r., n.m.r.).
(b) Loss of diene, formation of product (n.m.r.).
(c) Gas analysis, diene analysis, spectroscopy.

(d) Conductivity, u.v. spectroscopy of the ion. Less sensitive: f.p. depression.

(e) OH^- titration; spectroscopy, following u.v. spectrum of phenate ion.

(f) Gas estimation of N_2; possibly (but tedious) by gravimetric methods involving a diazo dye or by electrical conductivity changes.

2.2 Are H_o or J_o useful? Can Cl^- react with $HOCl$? How could one be certain of removing all Cl^- by complexing or by precipitation?

2.3 Sections 2.5 and 4.1.1.

2.4 $H_- = -\log a_{H^+} \cdot \gamma_{A^-}/\gamma_{HA}$. The ability of the solution to accept protons from HA (or, alternatively, to donate protons to A^-) would be affected by ionic strength effects, and it is also uncertain whether the activity coefficient ratio would be constant for all bases in the same solution.

2.5 The three series of acids fall into the group $Ar(CH_2)_n CO_2 H$ where n ranges from 0 to 2. Electronic effects between the ring and the carboxylic acid grouping would therefore be damped by transmission inductively through the aliphatic chain.

2.6 Is 9,10-dibromo-9,10-dihydrophenanthrene a precursor to 9-bromophenanthrene, or is it formed by a parallel pathway? Do the relative concentrations of the two products change with time, and can the adduct be decomposed to the bromophenanthrene *under the reaction conditions*? Can other adducts be isolated from aromatic substitution reactions? Is the rate of removal of the starting materials equal to the rate of formation of products? Is the reaction subject to base-catalysis?

Chapter 4

4.1 (c) The main electronic effect of p-OMe is a mesomeric effect. This is unlikely to be effectively transmitted through two carbon atoms. The intermediate must allow direct interaction between the aryl system and the carbonium ionic centre. The first intermediate does not show this; the last would imply the formation of m-methoxyphenyl products.

4.2 The rate of formation of RY will equal that of decomposition of RX ($= k_1[RX]$) only when all of the carbonium ionic intermediate forms product. In this case, a simple first-order plot results. As $[X^-]$ increases throughout the reaction, the diversion of R^+ to RX becomes significant. The rate at which R^+ is formed ($= k_1[RX]$) is therefore greater than the rate at which RY is formed; the apparent first-order plot (obtained by measuring the disappearance of RX, or the appearance of RY) therefore slopes downwards.

4.3 Approach of X^- along the direction of the $\overset{\delta+}{C}-\overset{\delta-}{Cl}$ dipole is the most energetically feasible way; approach along the $C-NR_3^+$ dipole is the least (section 4.2.1).

4.4 The solvolyses cannot involve carbanions, and so must be S_N1 processes (or mixed S_N1-S_N2 in the case of benzyl chloride). The halogen atoms apparently assist the stabilization of the carbonium ion.

4.5 The efficiency of neighbouring-group effects will be determined (i) by the proximity of the group and the carbonium ionic centre and (ii) by the stability of the ring formed.

Chapter 6

6.1 (a) Section 2.11.
(b) Is the rate depressed by added SO_2?
(c) Section 2.11.4.
(d) The initial rate equation is

$$\frac{dx}{dt} = k(a - x)(b - x) + k'\frac{(a - x)(b - x)}{x}$$

the solution of which is obtained by simple calculus (*J. Chem. Soc. (B)*, 1968, 712).

Chapter 7

7.1 Consider the predicted rate ratio if C—Cl bond-breaking were important in the rate-determining stage, and compare this with an accuracy of ± 2 per cent in the rate constant.

7.2 Relative to benzene, the aromatic system will be much more electropositive, and so each carbon atom will be more susceptible to attack.

7.3 Is *p*-nitroaniline an intermediate?
Is acetamide formed as a product?

7.4 (a) Benzyne intermediate!
(b) Nucleophilic attack by $NHOH^-$.

Chapter 8

8.3 The relative rates indicate an electrophilic process; the orientation of addition would determine the nature of the electrophile.

8.4 Solvation by HCl molecules, or increased acidity due to the formation of H_2SnCl_6, are possible explanations of the kinetics.

8.5 (a) The substituent effects indicate electrophilic attack in the slow stage of the process. The lack of solvent effect implies a concerted process in which external solvation is not significantly needed. See Lynch and Pausacker (1955).
(b) $RCO_2OH \rightleftharpoons HO^+ + RCO_2^-$ cannot be significant, or exchange would occur with added carboxylate ion.
(c) The formation of the protonated RCO_2OH molecule, not HO^+, seems likely.

8.6 (a) 2-Acetyl-1-methyl-2-chlorocyclopentane, which loses HCl rapidly to give 2-acetyl-1-methylcyclo-1-pentene.
(b) Me_2CHOMe (MeOH adds!)
(c) MeCHO and MeOH.

Chapter 9

9.2 No, the mechanism

$$RCOCl \xrightleftharpoons[\text{fast}]{\text{fast}} \overset{\displaystyle OH}{\underset{\displaystyle OH}{R-\overset{|}{\underset{|}{C}}-Cl}} \xrightarrow{\text{slow}} RC(OH)_2^+ + Cl^-$$

is equally appropriate, and has been favoured.

Chapter 10

10.2 Any initially formed amino-acid rapidly loses water irreversibly to form phenanthridone.

Chapter 12

In most cases, carbanions are generated; determine which carbanion is formed, and then see whether a nucleophilic addition, substitution, etc., is possible.

Index

Topics, rather than individual species, are listed here.